W9-BSB-680

MASTERS
OF
DECEPTION

ALSO BY MICHELLE SLATALLA AND JOSHUA QUITTNER

Mother's Day

Shoo-Fly Pie to Die

MASTERS OF DECEPTION

The Gang That Ruled Cyberspace

Michelle Slatalla
and Joshua Quittner

HarperPerennial
A Division of HarperCollins*Publishers*

A hardcover edition of this book was published in 1995 by HarperCollins
Publishers.

MASTERS OF DECEPTION. Copyright © 1995 by Michelle Slatalla and Joshua
Quittner. All rights reserved. Printed in the United States of America. No
part of this book may be used or reproduced in any manner whatsoever
without written permission except in the case of brief quotations embodied
in critical articles and reviews. For information address HarperCollins
Publishers, Inc., 10 East 53rd Street, New York, NY 10022.

HarperCollins books may be purchased for educational, business, or sales
promotional use. For information please write: Special Markets Department,
HarperCollins Publishers, Inc., 10 East 53rd Street, New York, NY 10022.

First HarperPerennial edition published 1996.

Designed by Caitlin Daniels

The Library of Congress has catalogued the hardcover edition as follows:

Slatalla, Michelle.
    Masters of deception : the gang that ruled cyberspace / Michelle
Slatalla and Joshua Quittner. — 1st ed.
        p.   cm.
    ISBN 0-06-017030-1
    1. Computer hackers—New York (N.Y.) 2. Computer crimes—
New York (N.Y.) 3. Computer security—New York (N.Y.) 4. Gangs—
New York (N.Y.) I. Quittner, Joshua. II. Title.
HV6773.25.N5S53   1994
364.1'68'097471—dc20   94-34417

ISBN 0-06-092694-5 (pbk.)

04   ❖/RRD   20 19 18 17 16 15 14 13 12

TO ZOE AND ELLA

ACKNOWLEDGMENTS

This book could not have been written without the help and support of friends and colleagues whose comments were invaluable. We thank our incomparable literary agent, Mary Evans, who was the first to see its potential. We also are grateful to our three editors—Craig Nelson, who liked the original idea, Charlotte Abbott, whose thoughtful editing vastly improved the manuscript, and Eamon Dolan, who shepherded us through the production process.

We also would like to thank Gary Cartwright, Bruce Sterling, John Perry Barlow, and Mike Godwin for guidance; and Peter Marks, Kinsey Wilson, and Robin Reisig for reading early drafts of the text.

MASTERS
OF
DECEPTION

PROLOGUE

You get a crazy-fast busy signal, like on Mother's Day when all the long-distance lines are jammed. Dial again. Busy. Dial again. Busybusybusybus—

Slam it down.

The phone system is crashing. The giant computers that route AT&T customers' long-distance calls, computers called switches, are shutting themselves down, behaving idiotically in a way that only computers can. Switches are the building blocks of the phone network, and when they start throwing electronic tantrums, the world stands still to watch. There's no choice.

It's happening today, all across America, as the phone company slogan "We're all connected" took on a new, darker meaning. Suddenly we're all disconnected. It's January 15, 1990, and millions of people are hopelessly jabbing the keypad, dialing the 1, dialing the three-digit area code, dialing the seven-digit number, waiting, getting a busy signal, trying again, getting a recording. One hundred fifty million long-distance calls a day travel these phone lines, threading through a massively intricate complex of earth-orbiting satellites, microwave transmitters, and millions of miles of fiber optic cable webbed across the globe. Thick-sheathed bundles of cable tunnel under the Atlantic Ocean. Long spindly strands are buried beneath railroad tracks or strung from utility poles that crisscross the Great

Plains. The signals they carry all converge on some of the most powerful and delicate computers on the planet. So much could go wrong, but AT&T has never failed so spectacularly before, never let its customers down. No natural disaster has ever compromised the phone company's dependability. A few weeks ago, an earthquake measuring 7.2 on the Richter scale collapsed freeways in California, and the phones still worked fine.

But not today. AT&T estimates that half of its customers' calls aren't getting through. Seventy-five million calls blocked. And the failure was sudden. At 2:25 P.M., one switch failed. Just one at first. Who knew why? It sent out a distress message to another switch, which should have been able to take over. But instead, the second switch went crazy, too, unable to cope with the bad news. It had a cascading effect, as one switch after another took itself out of service. And it was getting worse. Switch after switch after switch.

What caused this damage? What force is so powerful, so dangerous, so heedless, that it crashes a nation's communications system? In AT&T's New Jersey headquarters, dozens of technicians and bosses and engineers and control-center specialists all want the answer. They stare at the world maps, two stories tall, that line one wall in the control room. The maps suddenly flash red in unison, screaming emergency alert. No warning. No gradual escalation. One minute, the phones work. The next, parts of the Northeast are out. Businesses that depend on the phones are crippled, they're missing tens of millions of dollars in sales. Customer-service departments are sending their 800-line representatives home early; people are panicking because they can't get through to their elderly parents in other states; AT&T operators are giving out Sprint and MCI access numbers to frustrated customers; AT&T spokesmen are calling for calm.

And two teenagers in Queens are wondering if it's all their fault.

Mark keeps vampire hours, cramming-for-exam hours, Late-Late-Late Show hours. Hacker hours. In the middle of a Monday afternoon, Mark should be at his internship, which he has to complete unless he wants to drop out of yet another high school. If not at work, Mark should be asleep. Mark definitely should not be on the phone, calling Paul.

"Did you hear?" Mark says. "AT&T crashed."

"What?" says Paul. "What did you say?"

"AT&T crashed. It's on the news."

Paul sits upright on the red-and-black velvet couch in his mom's living room, startled awake from a nap, already knowing there's no way he'll ever go back to sleep. Not today, not tonight, probably not for the rest of his life. AT&T crashed. Oh God.

Paul's scared. Are they saying hackers did it? Did he do it? Could he have done it? He doesn't think so, doesn't want to believe it, has never in his life wanted to believe anything less. But the fact remains that the other night Paul was sitting at the computer in the basement of his family's house in Cambria Heights, the computer clandestinely hooked up through the phone line to a sensitive portion of AT&T's system.

Paul had diddled with the AT&T routing tables on a local switch owned by New York Telephone, typing first this command and then that one in an attempt to understand how the phone system works. This is what Paul likes to do. The routing tables are like train schedules that a New York Telephone switch uses every time a long-distance call comes in. The switch routes each call out over the proper carrier—AT&T, MCI, Sprint—so that the call can shoot instantaneously across the country and ring the phone in your grandma's house in California. It's a fabulously complex process, and no hacker in America could resist fooling with it if he knew how to get inside the system.

Of course, very few hackers could get past the security in the first place.

Paul had been sneaking in for months. He was the quiet programming genius behind the hacker gang known as the Masters of Deception. And so had Mark, the teacher in MOD, who knew more about phone company computers than just about anyone. A few of the other guys in the group might have been able to do it, too. Like Eli, the handsome smooth-talker. Or John, a newcomer and the only black kid in the gang. Or Julio, John's younger sidekick from the Bronx.

But who could be responsible for damage like this? No self-respecting computer hacker would ever destroy anything. No hacker would ever purposely hurt the phone system. Paul just wanted to look around. He just wanted to learn more. He'd know it if he'd done something bad. Wouldn't he?

And what about Mark? In the six months leading up to this crash, on Martin Luther King Day, Mark alone has made sixty-nine unauthorized calls to AT&T computers in Chicago and Portland, Maine. Each time, Mark secretly coupled his $300 computer to AT&T's multimillion-dollar monsters and, well, looked around. That's what Mark does. He's a hacker, and he breaks into unknown computers and tries to figure out how they work. His calls to AT&T computers totaled more than twelve intoxicating hours of exploration.

Paul and Mark know they're not allowed inside the AT&T system. It's illegal. The truth of the matter is that AT&T doesn't want some inexperienced teenagers from New York City testing the limits of its machinery. The kids are smart, there's no denying it. If they weren't smart, they couldn't get into the computers in the first place. And Paul and Mark are well intentioned. They just want to learn how things work. How else are they going to get access to study such a magnificent panoply of computer hardware? But once inside the corporate computers, Mark and Paul don't know what they're trampling, or where they're headed. Not really. They learn by trial and error.

"What happened?" Paul asks. He's afraid to hear the answer, afraid not to hear it.

"No one's saying," Mark says.

Paul pictures his friend at the cramped table that serves as his bedroom desk on the other side of Queens. Mark usually sits hunched over the beat-up telephone handset that he has modified and customized with much intuitive fiddling and electrical tape. Mark has taken the thing apart and put it back together, examining its innards so many times that the tape is like a truss holding in the entrails.

"I don't think it could be hackers," Paul says, in the same deliberate tone that he uses to measure out all his pronouncements. But what he is thinking, what he is fearing, what every bit of his brain is hoping is this: Please, don't let it be my fault.

The phone call from Mark to Paul lasts only a few minutes. In fact, if you really wanted to know how long it takes, you could ask the box. The DNR. The Hekemian Dial Number Recorder that sits on the twenty-third floor of the New York Telephone Company's headquarters in midtown Manhattan, down the hall from the cigarette smoke–filled office of the Security Department's toll fraud unit.

The box knows everything that matters. It watches Mark's phone, knows whenever someone lifts the receiver in the two-story brick row house in Elmhurst, Queens. The box records every phone number dialed from Mark's home, and it notes the hour: the start time, the end time, and the duration of each call. It funnels the information into a mini-computer that stores it for as long as anyone wants. And guess what? The box also watches Paul's phone. It's been watching for a while, months in fact. But don't worry. It's no big deal. The box doesn't stop them from making any calls.

The box just watches. And writes down what it sees. And waits.

ONE

It all started back in 1989, months before the AT&T crash, months before Paul and Mark even knew each other's names. The whole mess, which would grow into a world-class electronic gang war fought by hackers from New York City to Texas, started back before Paul even knew what a switch was.

If you want to trace it back to one night, to one single instant when you can say the whole story really began, you will see this image: Paul in the dark, peering into a garbage dumpster.

It's an early summer evening, warm on your skin, when Paul leans over the edge of the dumpster as far as he can. This is one way to become a computer hacker, the way Paul has chosen. He tries to snare one of the five or six invitingly swollen bags that sit in the bottom of the dumpster. It's not enough to be six feet tall, because Paul still can't reach the bags, not until his stomach becomes the fulcrum for his body and his feet actually leave solid earth. He dangles; the blood is rushing to his head, making him dizzy. And yet there's nowhere he'd rather be than here, rummaging around in this dark alley in a dumpster full of phone company trash, looking for computer printouts.

He came for the documents. But he also came for adventure. Right after scarfing down a quick supper, Paul had mumbled good-bye to his mom, who was still getting used to the fact that his dad had all of a sudden died. Then he hopped a bus to the end of the line. He got off in Jamaica, Queens, at Parsons Boulevard and Hillside Avenue, an intersection in a neighborhood hot with bodegas, bars, and beeper-rental joints. He stood for a while, afloat in the dwindling stream of commuters flowing from the subway. And then, a black Supra full of teenagers pulled up to the curb and the dark-eyed guy in the driver's seat checked him out, smiled. Paul got in.

"Let's go trashing," somebody in the back seat said. "There's a C.O. in Astoria." It's cool to be talking in a kind of hackers' code. The word *trashing* means climbing around in garbage, where you hope to find computer printouts that list secret passwords and logons. And C.O., as everybody in the Supra knows, means Central Office. As in New York Telephone's Central Office, in Astoria, Queens.

Somebody broke out the Ballantines and the Olde English 800s, and Paul took one. He knew only one of the other guys in the car, and that was his best friend, who calls himself Hac. Until tonight Paul had never physically met the driver, Eli Ladopoulos, only knew him by his nickname, Acid Phreak. This was not unusual. Everybody in the hacker underground has a handle, but you rarely get to meet face-to-face, and even more rarely get to sit together in a dark car speeding in the night toward a dumpster stuffed with telco secrets. With rush hour over, it took less than twenty minutes to drive from Eli's neighborhood to the other end of Queens.

When they reached Astoria, even the building itself thrilled Paul a little, this filthy redbrick giant that takes up most of the block. Across the front, chiseled over the door, it says TELEPHONE BUILDING. Like on a board game or something. Does the park across the street have a big sign that says PARK over it? No way. This building was named back when there was only one telephone company, Ma Bell.

They sidled up to the front of it. There are bars on all the two-story windows, and through them you can see a vast, loftlike, fluorescent-washed space. Imagine a library, hushed and eerie, but instead of books, all the shelves are filled from floor to ceiling with rack after rack of circuit boards. That's called the frame. There are even rolling ladders, with signs that say CAUTION: LOOK UP BEFORE CLIMBING. Like what kind of doofus wouldn't look up before climbing?

Upstairs in the building, connected to all those wires, is the switch. Paul doesn't even know what to call this large computer. But he's *seen* one now, and he'd soon learn its name.

The switch is the biggest computer you ever saw, and its job is to control every phone line in Astoria. When a phone company customer in an apartment over the Tae Kwon Do martial arts studio on 31st Street wants to order a pizza from around the corner, the phone call travels on copper cables to the switch, which funnels it to the pizza parlor's line. Multiply that by a couple of hundred thousand lines in this part of Queens, and millions of phone calls a day, and you see what we're dealing with.

Now, if you or your friends knew how to program a switch, or even knew a password to log on to a switch, you could start exploring. Go deep enough, learn your way around, and you'd be pretty powerful. Because then you could control everybody's phone service. You could create an unbillable number for yourself or your friends. You could listen in on phone calls. And best of all, you'd really know how this fabulously complex hunk of electronic circuitry works. That kind of omnipotence is beyond the reach of the CEO of New York Telephone. But maybe not beyond the reach of some teenage computer hacker who was tenacious enough to scrounge around in the garbage until he found a password.

This alone would be worth the trip. For a carload of teenage hackers, the opportunity to gawk through the big plate-glass windows at the matrix of electronic circuitry is better than Dorothy's arrival in the Emerald City.

Theoretically, tonight's conditions are ideal. It's an early summer weeknight, no rain, and the dumpsters are in a quiet alley on the side of the building. In the park, just enough kids are hanging out to provide an acceptable cover. There's plenty of promising garbage in big plastic bags, heavy as feed bags. It's much more auspicious than last week, when some guys took the trouble to climb the fence before they figured out the dumpster had already been emptied. Somebody on the street saw them, yelled up, "Hey, those windows don't open. You can't get any TVs up there." Like they would even want TVs.

And theoretically, tonight's crew is ideally suited to the job. These teenagers are not Central Casting's idea of computer nerds. Not a plastic pocket protector in the bunch, nobody squinting myopically through thick lenses. In fact, no one here wears glasses, and Paul and Hac, at least, are as muscled as the first-string running backs who graduated with them from high school last year. If they weren't so jumpy, they could toss around thirty-pound bags like Nerf balls. Look at Paul—he's the pale, serious one. He's always the quiet one in a crowd. Because he's a big kid, his silence is intimidating, whether he means it to be or not, as he stands staring with flat, Slavic eyes. Those eyes take in everything and return nothing.

Eli is his physical opposite. He's the one the girls like, the hip-hop guy, the cool one. Eli has a slow smile that starts like a conspiracy and spreads up to his eyes and pulls you in. His eyes are black as blueberries. His hair is as black as his eyes.

This is Paul's first time trashing, and frankly, if you knew him, you'd be shocked to see him here. He is, after all, the valedictorian of Thomas A. Edison High School's Class of 1988. Winner of the all-city computer programming competition. A boy with a future.

Paul learned to read before he got to kindergarten. He knew his colors and his numbers, and his favorite book was *Three Billy Goats Gruff*. He was captivated by the troll under the bridge. Maybe he would be all his life, maybe not, but at four, Paul was simultaneously drawn to and repelled by the powerful monster that lurked below. This was the authority figure that stood between you and where you wanted to go.

By the time Paul was six, he knew why he wanted to cross the bridge. He learned about the computer on the other side.

It looked like a typewriter but had extra keys. He saw it at his dad's office Christmas party in 1976. The winking cursor on the phosphorous screen enchanted him. How did it work? The boxy computer was one of twenty that sat, like a museum exhibit, behind a Plexiglas wall in the corner. Paul was taking the grand tour of the Royal Composing Room in Manhattan, where his dad was a typesetter; it was the first time he'd been to the office, and everyone was calling him Paulie, just like his dad did. When one of the guys saw him staring at the cursor, with that look in his eye, he took Paul and his father into the climate-controlled, dust-free sanctum where the computers hummed.

"Do you want to play around with one?" the guy asked.

Paul sat down at the keyboard, which was similar to his mom's manual typewriter. Using only his index fingers, he pecked in each letter of a short article from that day's newspaper, and watched as the words miraculously formed on the monitor. He could hit a key and just erase any error, type the word again. He typed the article perfectly.

The same guy came back, inspected Paul's work, and printed it. Out came a piece of plastic film, four inches by five inches, with the article, some story about the old JFK administration, typeset on it in black letters. Paul held it, still warm and smelling of chemical fixatives, and realized: If a computer did *that,* what can't such a wondrous machine accomplish?

It was software that made the computer work. It was the code, the precise and logical lines of instructions, that caused the silicon chips to react, just like neurons firing in a brain. It wasn't long before Paul was engrossed in magazines, reading about and actually digesting the rudiments of computer programming. He read *Byte,* and *Compute!,* and *Personal Computing.* He didn't have a computer of his own, and wouldn't for a few years, in fact he didn't even have one at school. But that didn't stop Paul from becoming a programmer. He learned to speak the programming language called BASIC. Writing programs without having a machine to run them is like learning chess moves without a chessboard. You have to hold everything in your mind. But here's the thing about Paul: one thought follows another logically in rapid-fire progression. Paul thinks like a computer, so it wasn't so hard for him to start writing for one.

The first program the eleven-year-old ever wrote was a challenge he read in a magazine: Devise a program that will search two groups of twenty numbers and tell which is the highest in each group. Paul thought about this for a while, thought about how it should work, with each command flowing from the previous one so that the problem was solved methodically. And then he wrote:

```
5 HI = 0
10 DIM A(20),B(20)
15 FOR I = 1 to 20
20 READ A(I)
```

```
25 IF A(I) > HI THEN HI =A(I)
30 NEXT I
35 PRINT "HIGHEST A VALUE IS "; HI
40 HI = 0
45 FOR I = 1 TO 20
50 READ B(I)
55 IF B(I) > HI THEN HI =B(I)
60 NEXT I
65 PRINT "HIGHEST B VALUE IS "; HI
70 END
100 DATA 5,8,2,15,7,3,7,8,36,18,
45,32,68,55,44,0,16,7,8,2
110 DATA 6,4,6,8,6,7,8,9,8,7,6,5,4,
3,2,14,15,33,22,11
```

The program would sift through the two sets of numbers at the end and identify 68 and 33 as the highest. Paul could tell just by looking at the terse lines of code that it would work. It had taken him ten minutes to write it. Then he wrote a program that tracked NFL football team statistics during the season: wins, losses, standings. Some programmers will tell you that it can ruin an up-and-coming hacker to learn to write in the computer language called BASIC because it's so clunky and primitive. Sure, you can write *any* program in BASIC—you could write code for the space shuttle if you wanted. But it would be millions of lines longer than writing the same program in a more efficient language. You can't express commands succinctly or elegantly in BASIC. Yet somehow the limitations didn't hinder Paul. The simple but sensible structure of the programming language hooked him. He wrote all his programs down in notebooks, page after page with very little erasure. The funny thing was, when he finally got his hands on a computer years later, he never checked to see if any of the programs worked. He just knew they would.

By the time he got to junior high and took his first real computer class, Paul could have been the teacher. A typical assignment took him about two minutes to complete. His teacher let him zip along at his own pace. The other students didn't know that BASIC was a simple interpreter for the even more complex native tongue computers speak, known as machine language. Machine language is a kind of numeric.

Morse code in which all the commands are expressed in a sequence of zeros and ones instead of in words recognizable to humans. Paul was by now talking directly to the computer in machine language.

Some people always remember their first car; others their first bike. These are the things that promise to liberate us from our ordinary lives and take us to places where anything is possible. Paul will always remember his first computer.

He got it in 1983, a gift from his parents, a Commodore 64 just like the ones at school. No one in his family knew anything about computers, but his uncle bought him Commodore's own *Programmer's Reference Guide*. Another guy might tinker endlessly with his '64 Mustang. Paul had his Commie 64, and man, did he get a kick out of peering under its hood. He'd open up the box and look at the microprocessors, understanding at the most basic level how the computer digested information, how the hardware interacted with software.

One thing about the Commie 64: it could run an awesome library of games, if you had the money to buy them. He bought Annihilator and 3-D Pac Man. They each cost twenty dollars. But Paul's taste got more expensive and the best games cost up to fifty dollars. So he traded with his friends. After a while, the games themselves weren't that challenging. Cracking their copyright protection was.

When the first generation of games hit the market, there was no such thing as software cracking. Games came on cassette tapes and were meant to be copied. But with the advent of floppy disks, the world changed. Software became a big business.

The software companies realized they needed to protect their franchises. The last thing they wanted was one teenager in Cambria Heights buying one copy of Zork and passing it out to one hundred pals. That's a hundred times fifty dollars the companies don't make.

So the companies started to lock down their software. It was easy at first. The programmers simply hid an intentional error on a part of the floppy disk. They called it Error 23. Then every time the game loaded, it had to verify that Error 23 was on the disk. The beauty of it was that most software programs wouldn't copy a disk that has an error on it. The floppy drive on the copying computer takes one scan of the disk, and says, "No way."

But wait a second. Didn't some kid just fork out fifty dollars for this program? Shouldn't he be able to copy it as many times as he wants? It's his, after all. What if his sister accidentally on purpose stuck his disk in the microwave? How's he going to play?

This was a widespread concern among teenage boys all over the country. It was perhaps their first conscious political stand. Even if they didn't know it, they were following a basic truth identified by *Whole Earth Review* founder Stewart Brand: Information wants to be free. To liberate it, these kids became "warez" dudes, amateur software pirates who put their collective ingenuity together. They traded tips for breaking lame copy protections. They even wrote little lockpicking programs, like Kwik Copy, that could copy a disk protected by measly Error 23.

It was a macho thing to do. Computer macho.

Naturally, the companies abandoned Error 23. Trying to be sequentially cryptic, they used Error 21 to protect their games. The first one with Error 21 was Flight Simulator, so you can imagine how anxious the dudes were to crack it.

Some guy in Canada wrote a program that could copy a disk with Error 21. He called it "Fast Hack 'Em."

That worked for a while. Then the hottest new game to hit the warez circuit was released: Summer Games, for Commodore. Its advent coincided with the 1984 Olympics, and it had eight different games, including a totally awesome pole-vault event that required you to jiggle your joystick as fast as you could to propel your little pole vaulter across the screen, up and over a little bar. The graphics were the best.

Paul did not have fifty dollars for the game. Now, Paul did not use warez-dude utility programs like "Fast Hack 'Em"—not that it would have helped on Summer Games. A guy who reads BASIC like it's his mother tongue can fast hack 'em on his own. If he wanted to copy a game, he simply loaded a program onto his computer that allowed him to X-ray the game. He could see all the underlying code on the game's disk, could see the hidden error, and tweeze it out like a splinter. That's why kids would bring Paul new games fresh out of the shrink wrap. Some of Paul's friends worked at computer stores, and at the end of their shifts, they could sort of borrow a game and bring it to him. The next day they'd return the original to the store. Paul never made a penny off this. He was more interested in defeating the troll.

Summer Games was tougher, though. Summer Games was the first one with Error 29.

Someone brought Paul a legitimate copy. At first it stumped him. Then he approached the problem like he approaches everything. He broke it down to its basic parts, figured out how each part worked separately, then figured out how they worked together. Just like an engineer. He scanned the sectors of the disk, looking for a clue. If he tried to delete the error, the program would still instruct the computer to look for the error and, not finding it, wouldn't run. The computer would say it found 0. Zero was a problem, because the software expected to find Error 29. But then, Paul figured it out. He could remove the error and at the same time convince the computer that it wasn't looking for an error.

Don't look for Error 29, he told the computer.

LOOK FOR 0

If you can't get rid of the troll, go around it.

It was a way of acknowledging that the error existed, while negating the power of the error.

Now he could make a million copies if he wanted.

Paul was proud of his work. He was an artist, and no artist wants to hide in a garret. Paul signed his work, inserting a new line at the bottom of his pirate floppies:

THIS GAME CRACKED BY: SCORPION PWS

Who was Scorpion? Paul didn't know. Scorpions were fast, silent, and dangerous. He knew who PWS was, fourteen-year-old Paul William Stira. He hoped the two were the same person.

Scorpion wouldn't be afraid to take risks. Scorpion wouldn't balk at climbing a fence and jumping into a phone company dumpster in Astoria. Which is not to say it's relaxing to be the "bag" man. Every time an ambulance screams by on its way to the Astoria General Hospital emergency room down the street, Paul starts and squints into the dark as if he expects to be pinned in the beam of a police flashlight. Hac's up on the roof of a utility shed next to the dumpster, and Eli is down on the street. Paul's the only one who might not make it to the Supra if the cops come. He tries to concentrate on the work: feel around in the dark for a bag, hope it's a dry one, then haul it up from the pit and pass it to Hac. It's like a bucket brigade, except that every time Paul hoists a bag, he almost loses his balance and falls into the dumpster. Do rats live in dumpsters? Paul wonders.

Scorpion mastered the art of cracking games. He needed new places to go, new worlds to conquer with his computer. By the time Paul entered high school, a friend had slipped him a sheet of paper with some phone numbers on it. The phone numbers were for hacker bulletin boards. They were electronic meeting places, these bulletin boards. All you needed was a modem that could connect you to electronic networks and a computer that could whisper a password, and you were in, reveling in the allure of the forbidden. Or so it seemed to teenage boys.

Your computer did the knocking for you. Just tell it what phone number to dial, type the password when you're prompted for one, and you'd be connected. Then you could type messages to other people who had logged in. People who liked the same things you did. People who might know a little bit more than you. People to hang out with.

At a time when boys in other parts of the country were out getting their driver's licenses, Paul was saving up to buy the one thing that stood between him and all the electronic bulletin boards in the world. A modem. A modem took the digital language of your computer and translated it into analog, the continuously varying sound waves that the telephone network was built to carry. Without a modem, Paul was just a hermit in a basement, typing alone. Without a modem, the phone numbers he carried around in his pocket were useless.

Any kid with a halfway decent computer setup and his own phone line for the modem (nothing pissed the folks off faster than having the house line tied up all night long) could procure the software to run his own bulletin board. The boards all had evocative names. You'd hear of a board called Pirates Cove, and you'd know it attracts software crackers who want to trade pirated games. You'd hear of a board called Phuc the Feds, and you'd know you're dealing with something else entirely.

These boards were real places, though some operated for less than a week before they imploded. Kids congregated on certain boards where they knew they'd find their friends. Sometimes they even arranged to meet at a certain time and go to other boards together. It was like an electronic street corner where they'd meet. Then cruise together. Through cyberspace. And nobody ever had to leave his bedroom.

A good board would also attract its share of hacker celebrities, kids who had built up reputations through their computer exploits, or by boasting of their prowess, or maybe just because they'd use a really clever hacker handle. A good board would attract "elite" hackers, the top tier of guys with really cool names like Erik Bloodaxe and Phiber Optik, guys who were in gangs, voted in because of their specialized expertise. There were wannabe gangs, and then there were real gangs, whose members would crow and scrawl their proud graffiti over bulletin boards throughout cyberspace. But no gang was more real, more revered, than the Legion of Doom.

The Legion of Doom was the best of the best from the fifty states.

A few of the Legion's members lived in New York. One LOD member, the brash and brilliant Phiber Optik (who arguably knew more about the intricacies of the phone system than anyone alive) was rumored to live in Queens. He cruised all the local boards, and if Phiber Optik graced a bulletin board with comments about this or that phone company secret, then other hackers spread the word: Phiber's on. This place is hot. A crowd would congregate. The phone lines would become busy. Hackers desperately calling, instructing their computers to redial, trying anything to get through, trying to get past a busy signal that was as implacable as any night-club bouncer.

You could find anything on a good board. A good board would be honeycombed with all sorts of treats in the form of text files you could read—recipes for homemade bombs, pilfered long-distance calling card codes, advice on how to cheat pay phones. Only the files weren't called "files." They were called *philes,* an homage to the phone. The telephone, that holy device, is the most important tool a hacker has, since it connects him to the biggest computer system in the world: the phone network.

So you can see why Paul needed a modem. He saved. It took a few months, lunch money here, a sixteenth birthday there. The

Master modem cost fifty dollars. It was the affordable modem, the VW Beetle of modems. It was slow, and it pulse-dialed like a rotary phone instead of using tones like a pushbutton. But it got the job done and it was the only one at Radio Shack in Paul's budget range. He also had to buy a phone jack and the cable to snake the phone line from the first floor down to the cellar, where the computer was set up. That stuff would cost between ten and twenty dollars; Paul wasn't exactly sure how much, so he didn't want to waste money on bus fare to the mall.

On a Sunday morning in April, his parents still in bed, Paul quietly slipped out the front door and breezed up 227th Street. Mock-Tudor townhouses all in a row. Cracks in the sidewalk. First daffodils in somebody's curbside garden. Cambria Heights was always held up as an integration success story; maybe because the whole process of one ethnic group moving out when another moved in was quieter here, less turbulent than in other parts of the city. Other Lithuanian families just like Paul's used to live in the neighborhood. In fact, before they died, Paul's grandparents lived within walking distance. But now a black family would move in, a white family would move out. After a few years, you realized that most of the white families with kids had sneaked over the border into Nassau County, Long Island. Soon you were the only white boy in your class at Public School 147Q. Of course, it was probably a lot easier to be the only white kid in a class in Queens than to be the only black kid in a class in Nassau County, and Paul, in his understated way, got along with everyone.

He turned left and walked another five blocks to reach an underpass beneath a rotting parkway that divided the city from suburbia, divided Queens from Long Island. Instantly, the roads were better paved, the houses spaced a little farther apart, the litter not as noticeable and graffiti nonexistent. And still he walked, south now, through Elmont and Valley Stream, a good three miles, so desperate was he to spend his sixty dollars and change.

Why did he want this so much? He didn't know. He couldn't answer that question, just like he couldn't tell you why he once painstakingly wrote out a list of the dozens of conversation forums offered by CompuServe, a computer information service that a friend showed him. There was no point to the list that took up pages in his noteback back then—it was like Columbus having a map before he even had a boat.

* * *

Paul got home from Green Acres Mall before his parents woke up. He had the thing hooked up before eleven that morning. It made a marvelous clicking noise when it dialed the phone.

He logged into a couple of bulletin boards, but he was just too excited to stay. He wanted to log in everywhere but was limited to the phone numbers on his list.

Within a week, Paul wrote a little program to help the modem find other computers. The program made the modem stay up all night after Paul went to sleep. The modem's job was to dial toll-free 800 numbers sequentially. It was a lot like in the movie *War Games,* a movie that influenced Paul and his friends in the same way that *Rebel without a Cause* had captivated an earlier generation of lost boys. In fact, this type of program is known as a War Games dialer. The modem might start by dialing 800-555-0000. If someone answered the phone, or if the phone just rang and rang, the modem would disconnect. Then it would dial 800-555-0001. And so on. The modem was hunting for other computers. When it found one, when the modem heard a fellow modem beep a greeting, Paul instructed his computer to make a note of that phone number.

All night long, Paul could hear the modem pulse-dialing. Clickety, clickety, click. It took a long time to scan numbers this way, but it didn't matter. Calling toll-free numbers doesn't cost anything.

The next morning, Paul would look over the night's work and see ten or even twenty computers' phone numbers on the list.

Within four years, he would have the numbers to a thousand computers.

During his junior year, Paul was a system operator of the high school electronic bulletin board. It was really more of a chat board, because nobody posted anything illegal, like credit card numbers, or sexy, like how to blow up the toilet. Hardly anybody called, in fact, except this guy who used the handle Hac, who also owned a Commie 64 and went to high school in Flushing. He lived near Shea Stadium, where the Mets play baseball. One night on the board Paul and Hac got into one of those arguments, where at the end you realize you're both saying the same thing. They argued over the definition of the word *hacker.*

First, Hac said that hackers didn't exist anymore. He said the word referred to the pioneering hardcore computer programmers of the 1960s. Those guys at MIT loved nothing more than writing, also called hacking, code. They were people whose love for computing bordered on the obsessive. They weren't guys who broke into computers, unless you count the times they sneaked processing time on the big mainframes, locked up in climate-controlled rooms.

Paul said he knew that. But he said hacking wasn't dead, pointing out that hackers today were interested in the same things only they cadged processing time by sneaking in over telephone lines. They were different from crackers, who just like to break into systems and have no idea of what to do once they get there.

It was then that they realized they held the same opinion. About a lot of things. So they met face-to-face. It's a funny thing about the computer world. You can talk to a guy for hours, have the most kick-ass conversations, and never know what the dude looks like.

They would hang out at the Queens Mall. Paul slouched through the shopping center, wearing his nylon "Black Ice" windbreaker and listening to Hac talk for hours. They sipped from really huge cups of mall coffee with lots of milk and sugar. Paul could drink four cups.

Paul would tell Hac about the new computers that his modem had found scanning. Hac, it turned out, had also been scanning 800 numbers. Paul told Hac that he explored new computers by calling back the number from his computer. He figured most of them were owned by private companies. Some would prompt: PASSWORD. Others wouldn't prompt him at all. They would just wait, passively, for a certain amount of time. Then they would disconnect him. Paul would try to figure out what sort of computers they were, and he would try to guess passwords. The computer system would identify itself as belonging to, say, the XYZ Corporation, and sometimes the password would be a variation of the company's name, like ZYX. Sometimes the password would simply be "test," or "guest," or "password." One out of thirty times, Paul could guess the password.

One day Paul sat at his desk randomly dialing telephone numbers. He was searching for a phone company computer. He knew from reading messages on bulletin boards that the phone company computer lines were often stashed up in the high numbers, up above 9900. Let's say Paul was looking in exchange 555. He would dial 555-9900, just to see what happened. Nothing. Then he would dial

555-9922, just for variety. He hated to dial numbers in sequence. He could dial any couplet, from 00 to 99, haphazardly, but without hitting any number twice.

555-9973. Nothing.

555-9918. Nothing.

555-9956. Nothing.

Then he hit one. 555-9940. And things got weird. He told Hac about it. Neither of them could make much of the thing.

Paul got a second phone line installed in his house and put up his own hacker bulletin board, called Beyond the Limit. The name came from a movie Paul saw advertised in *TV Guide*. He never watched the movie, just liked the name.

Not many hackers called Paul's board. In fact, only three did, and Hac was one of them. Hac got another kid he knew in Flushing to call, too. Hac was good at meeting people.

A year after they graduated from high school, Hac called Paul and told him he had met this incredibly elite dude who seemed to know a lot about hacking. The dude claimed a blind guy in the neighborhood had taught him to make free phone calls. Hac really wanted Paul to meet this guy. His name was Acid Phreak.

Skulking in the lamplight, Eli is supposed to watch out for telephone company security goons and for anyone else who might want to know why these scraggly-ass teenagers are climbing over an eight-foot fence. Down on the pavement, Eli can see the whole length of 30th Street.

It's crazy to climb over a fence onto private property when you're depending on a guy you hardly know as your lookout. If it had been anyone else but Acid Phreak, Paul probably wouldn't have done it. But there was something about Eli, this new kid, that made Paul want to be a part of whatever he's planning. Eli is a few months older than Paul, and even if he doesn't know as much about the mechanical aspects of computers and programming as Hac had been led to believe, he has other skills. He got his computer when he was fourteen, and ever since he's been calling people all over the world with his modem. Eli has friends everywhere. He's always game to try to break into different computer systems. He says he does it for "brags."

Paul hoists a bag over his shoulder, over his head, and hands it up to Hac. Then Hac hands it down to the sidewalk. That's the routine. But just as Hac's about to hand off the final bag, a man comes out of the telephone building and pauses a second longer than he should, and then gets into a car and rolls down the windows and just sits there. The boys freeze.

"What's he doing?" Paul whispers.

"I don't know. He's just sitting in his car."

Paul and Hac stand there, crazed alley cats, backs high, ears cupped, tensed on tiptoes.

And then the worst happens.

In the distance, they hear a siren. It's not an ambulance, whose aural signature Paul would recognize at this point in the evening. But it's definitely a siren, and it's getting louder. Closer. It's a banshee now, and it's just around the corner, and Paul, for one, has had it with dumpster diving. He climbs over the fence, as fast as he can, and follows a retreating Hac to the sidewalk. The siren's just about upon them, and they dash madly across the street, bags in tow, past the guy who's sitting in the car, now wide-eyed, watching the kids come leaping over the fence. Their sneakers hit the pavement with heavy slaps, and with barely a second to spare, they dive into a dark, safe spot in the park.

Just as a fire truck blazes past.

The siren is gone. They look at one another, their hearts pound. They can see the outline of the Triborough Bridge through the leafy trees. The green-and-white lights along its suspension beckon like a distant Ferris wheel, and it's an adventure again. They kneel on the handball court and rip the sacks open and paper printouts spill like entrails. The night is hot and the streets are hopping and you can probably even see stars. They don't look up.

TWO

Most of the dumpster's bags are full of garbage or containers of half-eaten Chinese food. But one bag is worth the trouble, because inside is a piece of paper that lists the phone numbers of about a dozen phone company computers in Queens. It's like a directory. No ordinary White Pages, though, this is a secret list of internal New York Telephone Company numbers. It's an excellent find. Now when you call up the phone company business office, pretending to be Lou in Provisioning, you can let it drop that you know all about a certain computer at a certain phone number. You're a part of the great Bell family, and of course the business office is going to give you the information you request. You can sound like you know the person: "Don't you remember me, Barry? We've talked before, fella."

Another great thing about the list is that it's like a schematic diagram that tells which computers are in Queens, and how they're related. How it all fits together.

The list is a map of a trip that, until now, Paul and Eli had been taking separately. Before they met tonight, they were just your typical lone spelunkers, rubbing their hands along cave walls in the dark. Now, with this shared experience, this set of directions, they're part of a team. There's a new breathlessness to them, a warm guy-feeling, because when you're trying to figure out something as labyrinthine as the nation's phone system, it helps to have friends.

The phone system is intentionally closed off to outsiders. Adolescent boys aren't supposed to be hooking up their computers to it, exploring its intricacies. But Paul and Eli and their friends are just playing an adventure game. This is merely a little trespassing. The very fact that they have to dive into a dumpster to get something as mundane as phone numbers for company computers only heightens their desire to get inside the system.

A few days ago, just before they decided to meet for the garbage run, there was a moment on the phone when Paul decided to give Eli something. He told him a secret.

Paul told Eli about the strange computer that answers the phone when he dials 555-9940, a number in Laurelton, Queens. "I think it has something to do with the phone system," Paul said, in a typically understated way.

In fact, Paul had found three separate phone numbers that all seemed to dial the same computer in Laurelton. Almost every day, Paul had been hooking up to this computer and trying to figure out what it does. The work was tedious.

At first, Paul would sit hour after hour at the keyboard in the basement, trying one combination of letters, then another, hoping to stumble across some command that the computer would respond to. He had plenty of time because Paul wasn't the type to do homework. Even though he was number one in his class, he did better when he didn't study too much. Studying could actually make him freeze up on a test, could confuse him when it came time to think through the questions and write down the answers. So forget that. Besides, you learn more from a computer than from any book. Computers interact with you. You do something. The computer responds. It's almost organic. So while his mom watched TV in the living room, he tried to hack the strange computer. When his brother went off to work on the night shift in the subway, he tried to hack it. It was like jiggling a handle on a door, wondering what was on the other side. He knew it was technically illegal. But who would know, and who was he hurting, and who could possibly care? He didn't think it was morally wrong.

555-9940
555-9941
555-9942

But each number seemed to lead to a separate place in the computer, because each one behaved differently, he told Eli. It was one of those absorbing puzzles that would ultimately teach him something. This was far more complicated than programming in BASIC; there was no manual to tell him how to proceed.

The last phone number was useless, actually, but it took him a long time to figure that out. Whenever he called 555-9942, he heard a sound, as if the computer were pulse-dialing, like one of those old rotary phones. He was calling the computer, and here it was calling someone else on a rotary phone.

Paul couldn't tell who the computer was calling, although all he had to do to figure it out was to count the clicks. But it dialed so fast he couldn't. He thought about this for a while, and then he dug out an old Panasonic cassette deck and recorded the clicks. He played it back, over and over, but still couldn't make out the number.

The problem was that at normal playback speed, it was still too fast. There must be some way to slow it down. There was no variable-speed switch on the cassette deck, like on a dictating machine. He couldn't put his finger on the capstan while it spun. What could he modify? He pulled out one of the four C batteries that ran the tape recorder, put it in backward to offset the voltage, and that slowed the speed.

After he figured out the phone number, he had his computer dial it. He was calling another computer in Jamaica, Queens, and he knew he connected because he heard the crash-and-bang of his modem colliding with another one. But his computer screen stayed black; whatever he was connected to refused to acknowledge him. He got no prompt on his screen. Nothing. He couldn't figure out what the computer wanted him to type. No combination of keys disturbed the blackness on his screen.

That happens. When you're trying to feel your way through the cave, sometimes you just walk into the wall.

Paul had better luck with 555-9940.

At least this number responded. The trouble was, the responses were wacky. The modem would connect, but when he'd try to communicate with the computer, any key he hit on his keyboard seemed to be the wrong key. The computer at the other end would send back to his screen this inscrutable message: S. He'd try again, this time dialing the 555-9941 number, and the computer would tell him: W. And that's what he'd get regardless of what he typed in response.

```
SSS
WWWWW
WW
```

He was trying every key combination.

```
WWWW
SS
SSSSSSSSSSSSSSSSS
```

One day in frustration he picked up his phone receiver and blew into the mouthpiece. A short airy blast, maybe three hundred milliseconds long. And now, his computer would sputter a string of bizarre characters across the screen. Paul figured it was some arcane timing pattern. Stranger still, now whenever he pressed a key, his own computer would ring a bell in response. Bing!

He realized that his susurrations had somehow reset the computer at the other end. Now he could get it to respond. He would blow into it again, and then a question mark would appear on the screen. Paul could tell it what he wanted to do:

```
? login
```

He was inside the phone system.

Paul experimented by typing various commands he knew from different operating systems he'd fooled with. No good. Then one day he typed "SHOW USERS," and boom—a list of authorized users appeared on his screen. He wrote down the names. The next time he logged in, he assumed one of their identities. Now he was an authorized user. He learned that if he typed "PRINT," and then the name of a file, he'd see a very helpful help file, which would explain what the directory was and what he could do with it. Like most computer systems, the phone company's computers were designed, after all, to help people. They were designed long before teenage hackers existed. They weren't built to repel a hacker menace once he had gotten into the system. So Paul got the computer to explain what the commands meant. For instance, by typing "PRINT QDN," he coaxed the computer to explain that the command QDN meant "query directory number."

He knew he was logged in to something powerful because one day when he was noodling around, he found a way to list phone numbers of people who lived in nearby Laurelton. Awesome. He looked for the phone number of a friend of his, a guy who worked at the *Village Voice*. He found it. But what could he do with it? He had a brilliant idea. He typed:

```
> QDN 5551234
```

Using QDN was a way to ask the computer to pull up the records of a particular phone number. Say the number in question was (718) 555-1234. The computer asked Paul if he wanted to place a "service order" to modify service:

```
SO:
```

Using the commands that Paul had found in the helpful help files, Paul typed:

```
> ADD ¢ 5551234 3WC ¢
```

Three-way calling, or 3WC, was now installed. Now when his friend wanted to add a third person to a phone conversation, he merely had to hit the flash hook on his phone. Paul could have just as easily typed "CWT" to add call-waiting to the line. The computer did the hard part. But Paul had all the power.

His friend never got billed for 3WC, and probably still has the feature to this day.

By the time he told Eli about some of this in the summer of 1989, Paul had been hooked into the phone company computer for about a year.

But what kind of a computer was it?

"It sounds like an SCCS," Eli said.

"What's that?" Paul asked.

"A Switching Control Center System," Eli said vaguely, something to centralize switches in one area. The explanation didn't satisfy Paul. Something didn't sound right. Of course, there was no reason it *would* be right. Eli had much less experience than Paul in actually penetrating the phone company's computers.

"An SCCS?" Paul repeated.

"Yeah, it's really big, and beyond my expertise," Eli said.

But that wasn't really a problem.

Even if Eli didn't know that much about hacking, he certainly knew a lot of hackers. He knew where to go for help. Because this is who Eli is: a kid who knows everybody, a kid who everybody likes, a kid who everybody wants to help. Paul, in the guarded way he warmed to anyone, had certainly come to like him just from talking on the phone. That's why Paul went along on the dumpster dive. That's why, a few days later, when they were talking about the mysteries of the Laurelton computer again, Paul went along with another suggestion Eli made.

Eli says they ought to find someone who would be interested in Paul's information, a phone company computer specialist who might be able to get inside other telco computers. Paul says OK.

Eli says, "I know this guy." But it turns out that Eli is not talking about just any guy. This is *the* dude. Eli's talking about *Phiber Optik,* says he's even encountered Phiber while roaming through cyberspace. Eli's never met him in person (but then, who has?), but Eli knows enough about Phiber Optik to know that he's the man with the answers. He's in the Legion of Doom, isn't he? He's the gang's *phone guy,* for God's sake. The Legion's exploits are legendary. The Legion is rumored to know how to break into ongoing phone calls. The Legion is rumored to have hidden the gang's own private bulletin boards inside corporate computer systems. The Legion's archives are rumored to be the repository for the best technical information in the underground.

Paul doesn't know anybody in the Legion of Doom, doesn't even know who's in this gang founded by a notorious hacker named Lex

Luthor. Eli says that if Phiber Optik got into the Legion of Doom, then Phiber Optik must be good.

You have to be a little brave to even suggest calling a guy like that. You have to be pretty sure of yourself, not afraid at all that the guy is going to hang up on you, or worse, listen to what you say and then *ridicule* you. You have to have a lot of confidence in yourself.

"Let's call him," Eli says.

Paul says okay.

"What do you want?" the voice demands. "I'm Phiber Optik of the LOD."

If you heard it, you'd think it was the Wizard of Oz himself, standing behind his curtain and making steam hiss and fires roar. Phiber Optik of the LOD. Both Paul and Eli hear it, the outrage in the deep voice that has answered the phone.

Now, Eli once "met" Phiber Optik on a bulletin board. But that's little comfort now, not with a real live member of the Legion of Doom on the other end of the phone, thundering and aggressive.

A guy like that doesn't like you, he can turn you into a toad—or at least turn your home phone into a pay phone. Every hacker has heard the stories, heard of some poor rodent whose mom picks up the phone in the kitchen to call Linda next door and instead of a reassuring tone hears the recording, "Please deposit twenty-five cents." Explain that to your mom.

Not that Phiber's response is totally unexpected. How does he know that he's not talking to a couple of lame wannabes on the phone? He gets these calls all the time. Ever since word spread that he's in the Legion of Doom.

What do Eli and Paul want of Phiber? It's obvious. They heard he was the dude who was into phone company switches. But that's the simple answer. They really want much more, don't they? They want him to teach them not only about the phone system, but also about all the sophisticated computers he's cracked, about the rare commands he can type, about the way his mind works. They want what any two boys with a little knowledge and a great curiosity want. They want a leader to show them the way.

They tell him they have access to a computer in Laurelton, and they want to know what it is, what it's capable of doing.

"We think we have this SCCS."

"You think or you know?" Phiber asks.

Hey, check it out, they say.

And then Paul, for the first time since he hacked into the computer, tells another living human the computer's phone numbers. Check it out.

So Phiber does.

He hangs up, plops down in front of the TV screen he has hooked up to his TRS-80 computer in his bedroom and dials away. It's a monastic setup for such an ambitious adventurer. A neatly made cot, a low bureau, a desk, a chair, makeshift bookshelves in the corner. The room is so small that the corners of everything touch.

This is where Phiber comes alive, though, in a Spartan room with one small window. Why would he need a bigger pane of glass? He has a view of the whole world from his computer. Outside this room, he's just another painfully thin adolescent with carefully combed hair, just another teenager colliding with uncertainty and the embarrassment of *being himself* every time his voice cracks, every time he walks into math class late, every time he says hi to a girl. But inside this room, Phiber is someone else. He's the smartest, coolest dude in cyberspace. He's intuitively hacking out the most complex programs and commands you can imagine. He's learning new things, going new places every day. By himself. He's *the* dude.

The funny thing about Phiber is, he's so far into the phone system that when he wants to hit a switch, he does it the hard way. He doesn't just dial the switch in question and connect. No, he logs in through something called the NYNEX Packet Switched Network. This network of computers is much more potent than any single switch. In fact, this network ties together every switch in the New York–New England telephone region. Each is one pearl on the necklace and Phiber has his hands on the clasp. But, ironically, he has never possessed a single specific phone number for any one of the switches.

Of course, he hasn't exactly been lusting for one. It's much more exciting to him to be able to envision the context, the vast splendor of a network of switches all working together, than to play around with any single component. As a little boy, he liked to assemble his own technical elements, but back then he wouldn't have recognized that the plastic models of ships he painstakingly constructed were really just networks of topsails and rigging and decks. The finished products were elaborate, though. And they all sailed.

Phiber may not possess the password to a switch. But he certainly knows a switch when he calls one. The Laurelton computer is a switch.

He phones Eli and Paul, and says, "It's not an SCCS. It's a DMS 100."

Phiber happily rattles on about how DMS stands for Digital Multiplex Switch, cheerfully informing them that the multimillion-dollar computer was manufactured by Northern Telecom, a giant telecommunications company that's one of the world's biggest vendors of phone equipment. Phiber loves to share what he knows with other serious hackers; he even was a kind of mentor to a kid named The Technician for a while. The Technician would call him up, ask a question, get an answer, hang up, try the hack himself, then call back for further instructions. It made Phiber feel like a real teacher.

DMS. The thing that Paul's been wandering around in for more than a year now has a name. And somebody's been there before who is only too pleased to show him the way. Because that's how it is with Phiber. He loves to share what he knows.

Phiber says, "You want to get together?"

And Paul thinks, Who is this guy?

He lives at the top of steep stairs in a redbrick row house just off Junction Boulevard in a part of Queens called Elmhurst. The neighborhood had been mostly Italian and Irish when Phiber's parents, Charles and Gloria Abene, moved in. Children of Italian immigrants, the Abenes had moved east from Brooklyn. As a young man, Charles Abene had dreamed of being a jazz saxophonist but ended up an officer in the school custodians' union. Gloria Abene worked in the billing office of the A&S department store in the Queens Mall. One luxury she allows herself every morning is a taxicab ride to work.

Just like Paul's neighborhood, Elmhurst over the years saw one generation of immigrants supplanted by another. This is the natural rhythm of New York City, where waves of newcomers wash in, and up, and out as the strivers move through the city on their way to the suburbs and assimilation. When Dominicans and Puerto Ricans began settling in Elmhurst, the Italian and Irish families headed east to Long Island. But a few hardy souls always seem to grab hold and refuse to be swept out of their neighborhoods. The Abenes stayed and raised three children in their long, narrow house. Their oldest son grew up, got married, and moved out. Their daughter grew up, went to college upstate, and moved out.

And now, the Abenes' precocious baby—who they know as Mark, not Phiber—is staying awake all night, sitting up at his computer until he sees what he calls "the stupidly sunny light" of dawn trip in through his window. Then he sleeps through math class.

Nobody could ever tell Mark what to do. He knew his ABCs before he was two and was trying to write words before he was old enough to properly grasp a pencil. He held it in his fist, and he shuffled down the steps one at a time, slowly, like the toddler he was, looking for paper. He remembers all this clearly, as if it happened yesterday, because there isn't much Mark hasn't consigned to memory. Phone company acronyms, obscure dialups, and being small enough to be bathed in the kitchen sink. It's all there, stored in his brain, and he can pull out the requisite memory anytime he wants to review it.

There's no point in suggesting that he go to sleep at a reasonable hour. Ever since he could walk, Mark has been a step ahead of the rest of the household. They've watched him in a kind of wonder, not sure where he's headed but trusting his sense of direction.

As a child, he was always asking how things worked. He liked to pry off the back of a radio to look at the parts inside, trying to see how they interacted. He took apart the cuckoo clock in the kitchen, the one that never worked, and he realigned the gears by intuition. He closed it up, and the bird started trilling. On the hour.

When he was in grade school, one of Mark's mom's friends gave him a dusty green book called *Using Electronics.* It was published back in the 1950s, and it came from a library that was selling off its obsolete books at garage-sale prices.

Armed with *Using Electronics,* he headed for Radio Shack. He wanted the parts to build a crystal radio. He asked the salesclerk for a condenser, and for a crystal diode. The clerk looked at him like he was nuts. The vocabulary had changed since the 1950s, when his guidebook had been published. What Mark needed, in the 1970s vernacular, was a capacitor, and a 1N34A germanium diode. That's the thing about electronics. The words might change, but the principles don't. Mark was learning to think like an engineer, breaking things down, seeing how they interact. He was approaching everything in the world—radios, cuckoo clocks, model ships—in the same way, as the sum of its parts, always reducing a system to its components, then envisioning how each unit fits together. There is an unassailable logic to this, a truth about how physical objects behave.

After school, in the afternoons, Mark would go down to A&S to wait for his mother to get off work. He hung out in the store's electronics department, acquainting himself with the first generation of home computers. A whole universe was on display, everything from the Apple II to the Timex Sinclair.

To Mark, each had a personality. The Apple was a rich kid, aloof and inaccessible. The Commodore was scrappy, but limited. But the TRS-80 was just like Mark. It was a smart little machine, elegant and able to do everything the richer ones could. The word that Mark used to describe the TRS-80 was *suave*.

You couldn't get a TRS-80 at A&S. The computer's initials stood for Tandy Radio Shack, that mecca of capacitors and diodes and crystal chips and breadboard for mapping out circuits. You couldn't get it at A&S, where his mom had a discount. But that's the computer he wanted. So his parents bought it for him for Christmas.

From the beginning, Mark saw himself as a scientist, and his computer was his most important tool. It was his probe, his means of connecting to systems that he wanted to examine. He did the electronic bulletin board scene, even used the ludicrous handle of Il Duce, after being impressed by a television program about the powerful Italian dictator. The best thing about bulletin boards was the clues the philes contained, clues about how to connect to the biggest, most complex system there was, the phone system.

The philes were tantalizing to a nascent hacker, because you knew that some of the mumbo-jumbo in them had to be true. They were written by hackers who had culled kernels of information from dumpster dives. The information, usually no good to them, was evidence of their exploits, and what better way to take credit than to type it up and post it for all to see?

Other philes taught Mark a new phrase: social engineering. Social engineering means tricking people into giving you information over the phone, usually by pretending you're someone they'd want to talk to. Because, as everyone knows, the best information comes from people, not computers.

The people most likely to give out information about how the phone system works are the helpful staffers in the phone company's own business office. The public calls all day with questions, legitimate questions about their bills and their service. The office also handles calls from field workers, the guys stuck up on a telephone pole, the guys who make sure that each customer's phone line works. The business office staff is used to giving out detailed information, politely.

Mark started connecting through the business office. Of course, he couldn't go there physically, because then everyone would see that he was in junior high school and obviously not a lineman at all.

So he went to a pay phone down the block from his house one afternoon and from there dialed the business office. The ambient street sounds only supported his claim that he was up on a pole.

A woman answered. "Business office."

Mark said, "Hi, I'm calling from the Repair Service Bureau. My name is Joe Linerman, and I don't have a directory handy, and I need a number for line assignment."

Who wouldn't trust a voice as deep as that?

The woman gave him the phone number for the line assignment office.

So Mark immediately called line assignment, and told the voice at the other end that he needed the cable-and-pair number for his own telephone number. Dutifully, the voice on the other end read off two long, hyphenated numbers that designate the specific wires that lead into and out of the Abene household.

Think of the "pair" as a tributary and the "cable" as the river, all of it flowing into the phone network. The numbers designating a specific cable and pair are simply one unit of information. But if you amass more units, you can start to understand how the system's components fit together. If you can call someone out of the blue, and she helpfully tells you the secret numbers that New York Telephone uses to identify someone's phone service, what won't she tell you?

Mark is patient, and he is a scientist. Nearly everyday, he called back and tried to compile more units of information. He learned that the business office has access to a computer system known as ICRIS, which stands for the Integrated Customer Record Information System. They have to be able to use that to field customers' billing inquiries. From there, he learned about LDMCs and PREMIS and PREMLAC and LMOS, all acronyms for systems that administer and map the sprawling telephone network.

In time, Mark learned how the wire pairs that run into your house are connected to cables and how the cables are connected to trunks, and how all of it is controlled by the switch. Mark the scientist didn't read this in a phile somewhere, he learned it through trial and error, through theory and experimentation. The knowledge he assembled is as rock-solid as anyone outside of New York Telephone could hope to have. When he was in the system, he had this uncanny sense of the architecture, as if he were physically inside, a sightseer in a cathedral.

Mark the teacher needed to share the knowledge that Mark the scientist accumulated. The usual bulletin boards were a bore, it was pointless crowing to wannabe hackers, kids who just follow a recipe for hacking without understanding why you include certain ingredients. Messages posted on bulletin boards are stripped of inflection and facial expressions, and can hit a reader full in the face. The ideas and words are not softened by the physical signals that a speaker uses to convey that he's trying to be funny or silly or just plain honest. If you've read it all before, if you know more than the average kid, then scrolling through screen after screen of adolescent ramblings and posturings can be, well, annoying.

Mark was looking for a more educated audience, and one day, he got a promising phone number and password from a Florida hacker named CompuPhreak.

Catch-22. That was the name of the elite, super-secure bulletin board that belonged to the phone number and password. The board operated in Massachusetts, and it was rumored to be a hangout for members of the most notorious gang in cyberspace.

Mark had stumbled into the clubhouse of the Legion of Doom.

The Legion's founder, Lex Luthor, was so well known in the underground that his face could appear on dollar bills. Lex expropriated the gang's name from the band of arch criminals that plagued Superman

on the Saturday cartoon show. It was a sign of the times that Lex Luthor
was a child of the small screen, more familiar with the cartoon show
than with the comic book.

Catch 22 was a sneaky board.

It had Lex Luthor's patented Big Business Computer Response.
In other words, when you called it, it would appear as if you had
dialed into a huge mainframe.

ENTER CLASS:

would appear on the screen. The only acceptable response would be
to type VAXB. If you typed anything else, the system would
disconnect you. Cool, huh?

The password was changed by word of mouth every few months.
Another thing Lex Luthor insisted on was that no one would post
codes, which are pilfered calling card numbers. That sort of petty
crime was beneath LOD and offended the sensibilities of many
members.

Mark logged in, using the password CompuPhreak gave him. He
quickly moved to the phreak part of the board, the place where
hackers discuss technical intricacies of the phone computers they're
trying to crack. Mark took one look and practically laughed. Here
was the vaunted Legion of Doom and they had nothing better to post
than a few tired telco philes that had already made the rounds.

He started posting. He explained one acronym. Then another. The
thing that immediately set his postings apart from everyone else's
was that his were so obviously correct. Here was a guy who appeared
to understand the whole system—central office to junction box and
every multiplexer in between.

Word spread to Lex Luthor himself. Check out this new dude.
The telco wizard, Phiber Optik. Where did he come from? How does
one become a member of LOD? It wasn't like you had to prick your
finger and swap blood with Lex's protégé, Erik Bloodaxe, or
anything. Gang members on the electronic frontier don't live in the
same states, wouldn't recognize each other if they were standing
shoulder to shoulder on the bus. In fact, Mark swears he never spoke
to Bloodaxe before joining LOD. But the way Bloodaxe remembers

it, the encounter occurred like this: One day down in Texas, where he was a college student, Bloodaxe noticed that Mark had started signing his postings "Phiber Optik of the LOD." And Bloodaxe thought, who is this kid?

He immediately phoned north.

"Hi, is this Mark?"

"Yeah."

"This is Chris—Erik Bloodaxe," said Bloodaxe, whose real name is Chris Goggans. "Why in the hell are you signing your name LOD? You're not in LOD."

Mark thought for a second, then said solidly, "I'm in LOD."

"No one is in LOD unless we all vote on it," corrected Chris, who explained the rules. A unanimous vote is necessary.

Then, for some reason, the tone of the conversation shifted to what both teenagers really cared about. Hacking the phone company. And Chris realized that Mark really did know as much as people had been saying, maybe more. This guy was good.

The actual vote came a few weeks later. Mark was in.

Just like any schoolyard pack of boys born in the shadow of *The Dirty Dozen,* "Hogan's Heroes," and "Mission: Impossible," the LOD members all fancied themselves specialists in some dark art. One kid might know how to make a wicked blue box, a device cobbled together from top-secret Radio Shack parts that simulated the tones of coins dropping into a pay phone. Another might be an expert in programming BASIC.

And Mark? He could trace the route of a phone call, from New York to Paris, recalling in loving technobabble each photonic hop. He could describe, in detail, the different kinds of computers that run different aspects of the phone company's business. He knew the meaning of the phone system's every English language–mangling acronym: MIZAR, COSMOS, SAG, LMOS. He could explain it to anyone. Indeed, he loved to, in eye-glazing, brain-fogging, soporific detail.

And he had just turned seventeen.

To tell you the truth, a few members got a little sick of the new prodigy. He was brash and had what some out-of-state members recognized as a New York attitude. And he didn't give a rat's ass who thought so.

He seemed to revel in belittling blustery hackers who posted misinformation. He loved nothing better than trapping some nitwit who thought COSMOS was some double-secret key to the phone company kingdom. (Duh, the name sure sounds important, doesn't it?)

People were starting to notice. Like Chris. One day he and a friend, Dr. Who, were hanging out on a hot bulletin board called The Phoenix Project. Who did they run across but Phiber Optik of the LOD, eviscerating some poor pretender.

And this was what Chris thought about Mark: a real arrogant, smart-ass punk.

tHREE

It's a hot July day, nighttime already, when Eli cruises up to Mark's house in his black Supra.

As Mark comes down off the stoop, before he even gets in the car, it's clear he looks nothing like his megabass voice would lead you to believe. He's as thin and pale as the underside of an index finger and has otter-sleek black hair, meticulously clipped and combed. He styles it himself, doesn't trust a barber to preserve the precise geometric layering. His jeans look like they're just about John Candy's size, so they're beyond baggy on Mark, cinched with a thick belt and pooling over his shoes. His shirt is clean and pressed, his pants are clean and rumpled, his hair is clean and shiny. He gives off a good smell as he gets into the car, a fresh, soapy fragrance that fills a space and makes a car owner self-conscious about all the crumbs on the floor and all the dust motes on the dashboard.

Eli says Paul is hanging out at the Continental, the outdoor strip of stores at 71st Avenue and Continental Boulevard, so they drive over there and pick him up. And now, of course, the evening's main activities can commence.

You might call it hanging out. Eli calls it "The Mission." Maybe they barely know one another, but already they've got a bond. Imagine meeting the only other two people in the world who think exactly like you, who have totally the same goals in life, who would rather hack into a phone company computer than do anything else you could suggest. And they live in Queens.

Maybe it *is* a mission.

They cut across the borough, and end up at Eli's house in Jamaica. They don't see his mom, even though she should be home from her job as a receptionist by now, so they don't have to endure the exquisite embarrassment of encountering her—and worse! maybe having to say hello—before they get to the privacy of Eli's room. Eli's dad is a cook, and his parents don't get along. A lot of people think they might get divorced.

Eli's computer is set up in his bedroom, a room that screams "TEENAGE BOY LIVES HERE," with a life-size poster of a bikini-clad model on the closet door, a white cordless phone, a second desk phone with two lines, a TV and cable box. Eli has a York cassette recorder, a Spectron telephone speaker amplifier, and a shoe box containing 120 floppy disks. The bed is made, the dresser is dusted, the clothes are put away. If Eli had a refrigerator in here, he might never have to come out of this room. Eli walks to the computer table and flips the switch on his Commodore 128. The machine whirs to life.

Mark loves to explain, and Paul and Eli crowd around him at the monitor. As they log into the Laurelton switch to start exploring, he describes every command they're typing—even the commands they already know—in precise, easy-to-understand language. He knows everything. And Mark is just as excited by this session as they are, because he senses that finally he's met two other hackers who can ride at his pace. For his part, Mark will always think of this evening as "a meeting of the minds." They forgot who they were, and where they were, and thought only about where they were headed.

Mark has shown them how to use the NYNEX Packet Switched Network to jump off into other switches as well, and tonight they traipse around in the Hollis switch system for a while. In earlier phone conversations, Mark has told them different ways he's found to get into phone company computers, and Paul took it all in. So tonight Mark never has to repeat a phone number, never has to explain the meaning of a command to Paul. Mark types it and Paul absorbs it, because the progression of commands on the monitor is distinctly logical. Paul watches just once, and the new numbers are committed to memory. An hour later, when the three boys want to return to a computer they'd said good-bye to, Paul is at the keyboard and competently types in the commands again, no misses, no questions.

Imagine the feeling. Every pathetic BBS is filled with philes on how to crack this computer system and how to crack that one. Anyone but a moron knows that it's baloney, a farce, techno-silliness that doesn't work. Any kid with a modem talks about hacking the phone system in the same way that any teenager talks about getting laid. In other words, it happens. But rarely—and usually to someone else.

But now, here they are, typing on a piece of $300 equipment, hooked into what seems like one of the mightiest computers in the world. For someone else, it might have all sorts of catastrophic appeal. You could do anything, even cut off phone service to the whole Laurelton neighborhood. But that's anathema to them; they'd no sooner crash a computer system than they would cut off a finger. That's what they tell each other. They believe in the hacker ethic: Thou shalt not destroy. It's OK to look around, but don't hurt anything. It's good enough just to be here.

It's late now, the mission has turned into an all-nighter, and it's the bold hour when all the authority figures they've ever known are already asleep, oblivious to the escalation of the shared kinetic energy in this room.

They log in to one of New York Telephone's COSMOS computers, whose intricacies Mark is happy to explain. COSMOS is a grand-sounding acronym that turns up on any self-respecting hacker BBS. Phile after phile is written about it, as if the system with the fancy name was somehow the very key to the Bell System. It's all nonsense, explains Mark: COSMOS, he says, stands for Computer System for Mainframe Operations, and is nothing more than a giant database of work orders. It's an operations system that phone company employees use when they have to change something on a phone line. COSMOS has a directory of customers' phone numbers, archives that list the number of the cable and pair that run down from the big silver box on the telephone pole outside and into your house. You can look up anybody's phone service on COSMOS. Just read the code line for somebody's service, and you suddenly know such intimate details as whether the customer has three-way calling, or call waiting, or call forwarding. That's what it is, Mark explains. Nothing to it.

Mark shows them how to call up actual service orders for the phone lines that lead into each of their own houses. It's so exciting seeing it there on the screen. They're doing this with a laughably slow modem, a modem that pumps data at the rate of 1200 bits of data per second. It takes forever for the screen to refresh itself at that rate, but it's oh so sweet when it happens. They see Eli's own phone number. They see the features he has on his phone. The boys have an uncontrollable urge to crow, to scrawl graffiti across this privileged line of computer code as it blinks on the screen. If any one of the boys were hacking alone, in his bedroom, he wouldn't feel the same way. But here they are, together, and they need to mark their shared journey. Thou shalt not destroy, no, of course not. But they traveled here, this is their turf, and they want to plant a flag. That wouldn't be destroying anything, would it? And COSMOS is inviting them to do just that, with its tantalizing prompt:

```
JA%
```

They decide to write Eli's hacker handle on his phone line! Right there, right on the computer! Just write it in. They know the commands to type. Mark was the first to figure them out. They tell the computer to execute a service order:

```
JA% SOE
```

Then they tell the computer that the service order will modify service on Eli's phone number (let's say his number is 555-9365):

```
_I TN=555-9365
```

Then they tell the computer to execute the order on the next day:

```
_H ORD=C1AP1234,OT=CH,DD=07-13-89,FDD=07-
13-89
```

Then they tell the computer to add remarks to Eli's phone number:

```
_I RMKT=ACIDPHREAK
```

That's the end of the service order, they tell the computer:

```
_E
```

It's dizzying, the risk, because what if some phone company employee calls it up the next day and sees "ACID PHREAK" written into the code? On the other hand, why would anyone call it up ever, unless the Ladopoulos family requests a change in service? The boys leave it. They aren't hurting anything.

They write "PHIBER OPTIK" on Mark's phone line. They leave it that way.

By the time Eli drives them home, in the late-late part of a night that's ready to become morning, they are talking about what they'll do when they log in again in a few hours. There are so many places to go, so many things to learn. Is it possible to listen in on ongoing phone conversations? Is there a way to get into the phone company's automated message accounting, which contains billing information and lists the phone numbers that a certain customer calls?

By the time the night ends, all three know one thing: They can't wait to get together again.

There's a big street map of Queens on one wall in Eli's room. From the East River to Long Island, the map details the hundreds and hundreds of crisscrossing streets and avenues and boulevards and highways that cut up the borough. It identifies the parks, the airports, and the jail at Rikers Island.

Stand in front of the map, and trace the grid with a finger, and you could see how Eli's house was the natural hub. It was smack in the center of all of Queens, with Mark's redbrick house at the end of a spoke on the northwest and Paul's frame house at the end of another spoke that stretches southeast. Not only is Eli's house positioned perfectly for a command center, but his Commodore 128 was more powerful than the computers that Mark and Paul owned. Mark never went to Paul's house, except once to drop him off, and then he certainly never set foot inside. Paul sometimes went to Mark's house, but there was usually nothing good to eat. Mark wasn't interested in snacks, so he didn't offer any. One time Paul and Hac brought their own dinner, lasagna in foil, and they ate on Mark's front stoop. Just to make a point.

A lot of times, the guys got together at the Queens Mall, where Paul liked his coffee sweet, and Mark liked to get mashed potatoes from the Kentucky Fried Chicken outlet. He was queasy a lot, stuck to bland food, and rarely ate before the afternoon. It was just part of who he is. The problem with real mashed potatoes is that people make them lumpy, but he could eat several portions of those reconstituted, processed starchy ones at the mall. Go figure.

But how often could you go to the mall? There was no computer there.

So Eli's house, Eli's bedroom, that was the place. It was the closest thing to a clubhouse that they'd ever have. Four or five guys could hang out comfortably in that room, on the bed, on the chair, on the floor, on the computer, drinking caffeine in any of its splendid forms, running the modem off one phone line while keeping the second line free for voice calls.

They were learning a lot about the phone system. From the inside. If Eli called it "The Mission," Mark thought of it as "The Project." And Paul? He just wanted to know more. Even when they were at home, hacking on their respective computers, they kept in contact, phoning two or three times a day, updating each other on what they were finding in phone company computers. What does this command do? What does this acronym mean? Often, though, they ended up working together, at least by evening.

They knew so much now, collectively, so much more than they did on their own. But nobody knew they knew. It was frustrating being so omnipotent. Imagine being able to fly. Imagine being invisible, and now imagine not being able to tell anyone about it.

Then one night, about eight o'clock, they got an idea. The three of them were hanging in Eli's room, looking at the map of Queens on the wall, and realized there was at least one place in the borough that the three of them could fly to.

It was called Anarchy.

It was a virtual neighborhood, actually, this computer bulletin board called Anarchy. Some kid ran it from his bedroom in "Outer Queens," as Mark calls it.

The boys in Eli's room had some unfinished business with Anarchy.

Paul had discovered the phone number for the Anarchy board once when he was on a different board. Just for the fun of it, Paul had even signed up as a user on the board, which had a special section for posting philes specific to hacking and phreaking (hacking the phone system). And one time when Paul was at Eli's they had all logged in to take a quick tour of the hack/phreak archives. The philes were garbage. One phile said that COSMOS ran the whole phone system! Another example—a bunch of philes blathered about REMOBS, the actual physical units that attached a pair of phone line cables to allow remote access. Well, the lamers on Anarchy bought into the stupid misconception that if you possessed some magic phone number, you could dial it to access REMOBS, then key in a three-digit code and someone else's phone number—and *eavesdrop* on conversations. That might have been true in a parallel universe, but in the Bell system? Please.

The boys in Eli's room had tried to set things straight, posting a correction on Anarchy: YOU'RE WRONG. But that just started a flame war, with all the board's users joining in to defend the status quo. It was a pushing and shoving match, a schoolyard brawl that the boys in Eli's room had never really forgotten.

The kid who ran Anarchy called himself The Graduate. He was the kind of puffed-up wannabe that Mark calls a "larva hacker," a kid who says "I specialize in networks," as if he really knows the first thing about networks.

It burned, if you want to know the truth. Here were Mark, Paul, and Eli, putting in the hours, doing the real work, truly figuring out how the System works. Then along came some kid like The Graduate, some poser type who knew nothing, and the kid was spewing online, beating his chest, and giving everyone a headache.

What could you do about it?

Actually, the boys could do a lot, knowing what they did now.

It so happened that the Anarchy phone number was archived on a switch that the boys in Eli's room could control.

Maybe what happened next was inevitable, maybe not, but one thing was for sure. There was no turning back.

"Let's take over Anarchy," someone said.

Maybe there was a split second when one of them could have demurred. Probably not. And besides, no one did. A group mind had already taken over. Something bigger than all of them had been born.

The first phone call the boys make in their scheme to overthrow Anarchy is to a company—we'll call it PhoneBox—that operates a voice-mail system in Manhattan.

This is how PhoneBox does business: it rents out voice-mail boxes, which are kind of like post office boxes for phone messages. PhoneBox has dozens of phone lines assigned to it, which customers rent. The customer's business associates and friends can then leave messages in the voice-mail box. All the numbers that PhoneBox rents to customers begin with the Manhattan area code 212, followed by the prefix 333. The last four digits, the extension, are unique to each customer.

Places like PhoneBox promise customers anonymity and therefore attract a certain clientele. There's an odor about the place that hackers find irresistible, like a carnival midway where all sorts of sleazy transactions are consummated in the back alleys. The boys in Eli's room would hack a customer's line (the default password is the same as the last four digits of the phone number or sometimes there'd be no password at all—smart) and listen to private messages. They would hear nameless voices leave what sounded like stolen calling card numbers.

It takes the boys in Eli's room no time to set up their own voice-mail box on a vacant line in the PhoneBox system. They record a genial greeting on the PhoneBox answering machine.

Next, they log in to the phone company switch on which The Graduate is a customer. On the screen they call up the phone number for the Anarchy bulletin board, and add a feature to its phone service—call-forwarding. Then they forward all the bulletin board's calls to the answering machine on the PhoneBox line.

It's a fine hack.

Now any kid who tries to log in to Anarchy will unknowingly

bypass the bulletin board. The phone call will be routed automatically to the PhoneBox voice-mail box, where a voice will greet the caller with this message: "Hi, this is The Graduate. The board crashed, and I lost all the files. If you want to keep your account, leave your login and your password now, and we'll set it up for you as soon as the system is up again." The message will pipe through the little speaker on the computer. "Pick up the phone, and leave your password now." Beautiful.

Of course, the calls come streaming in, one after another, and every caller leaves a message. Nobody can get through to the real Anarchy board, which sits dormant, as neglected as an unplugged vending machine.

Everything's cool, the boys are culling all the logins and passwords they could possibly want, listening to each voice message and laughing uproariously as the callers leave their private information on the machine.

The boys are hacking hackers.

Then, one of the messages on the voice-mail box gets their attention. The message has a different tone to it. Worried, confused. Something's wrong. The caller isn't your everyday lame hacker, no, he's the co-system operator for the whole Anarchy bulletin board. He's The Graduate's best friend, the only one entrusted with running the system when The Graduate is out of town. Which it turns out he is, visiting his grandmother in New Jersey.

The co-system operator must believe that the system really crashed and that The Graduate is trying to patch it back together, because he leaves his login and password. Too funny! Doesn't even pick up on the fact that the voice claiming to be The Graduate is an impostor, but hey, that's voice mail for you.

The boys in Eli's room decide to bypass the voice-mail box now, they're having such a good time. They change the call-forwarding instructions, so that the calls are forwarded directly to Eli's house. Eli's answering the phone himself, live, taking down everybody's login and password, and the calls keep coming in over the white cordless phone.

But then the co-system operator calls back. It must have really rankled him, or maybe just seemed strange, because he's on the line again and you can tell he's suspicious.

He hangs up. The phone rings, almost immediately.

Paul answers this time, and he starts to say, "Hello, this is The Graduate—"

But then the caller breaks in.

"I'm The Graduate," the caller says.

It is The Graduate himself, no lie, calling from New Jersey to try to figure out what went wrong with his bulletin board. The co-system operator had given him the heads up. But The Graduate thinks he's calling a phone line that leads to his own bedroom. That's the number he dialed and he has no idea that anyone has forwarded the number to another location.

So what The Graduate thinks is: Who's standing in my bedroom, in Outer Queens, rifling through *my* disks and *my* notebooks and *my* hard drive? And who knows, maybe even *my* underwear drawer?

"Who is this?" The Graduate demands from New Jersey. "What are you doing at my house?"

Paul, whose strengths do not lie in verbal volleying, shoots the phone over to Eli.

Without missing a beat, Eli says into the receiver, using a deadly serious voice: "We're the Secret Service special task force and we're taking away all your computer equipment."

My, my, my. You can practically hear the thud when The Graduate's heart falls right on Granny's rug. Because, as every teenage hacker in America knows, this is how a promising young man gets cut down in the prime of his underground BBS life. The Secret Service busts you. It's well known that the Service handles investigations of electronic intrusions. And every hacker in America fancies himself a clever intruder. The Graduate is no exception to this conceit, and, all the laughter in the background of Eli's end of the conversation notwithstanding, The Graduate does what any kid would do confronted with the dreadful certainty that the jig, finally, is up.

He panics.

He's being raided! He's headed to jail, where he will be forced, no doubt, to be another man's wife!

Worse, what will his parents say?

"You're in a lot of trouble," Eli says.

"I didn't do anything wrong!" The Graduate yells. But not convincingly. "What do you want?"

The Graduate will do anything to make this agony end. There's a hitch in his voice, he's near hysteria.

"We're packing up your stuff, and your parents aren't going to like this," Eli says. "Don't worry. We'll call you."

The boys are listening in, and somebody yells, "Hey! Let's see what's in his refrigerator while we're here!" But apparently The Graduate doesn't hear that part, or if he does, he assumes that refrigerator searches are standard procedure, because he says, "OK, OK, OK."

And then, just like that, he volunteers his own login, and his own password, which is—believe it or not—SUPRA, like Eli's car. Since The Graduate is the system operator; his password gives the boys in Eli's room the power to do anything they want with his bulletin board. Change things, rewrite programs, delete philes, you name it.

After they hang up, the boys in Eli's room start to pack up. They erase the message on the voice-mail box. They erase the call-forwarding command on The Graduate's phone line, and they also give Anarchy a new phone number that hasn't been assigned to another New York Telephone customer: (718) 555-0000. That's to keep anyone from calling and tying up the line.

Then Eli calls the new phone number, types "SUPRA," and starts looking around in the guts of the Anarchy system.

The boys are looking over his shoulder, shouting suggestions, having a good time, when something happens.

PLIK

This weird word appears on Eli's monitor. If it's an acronym, even Mark has never heard of it. Plik! Eli tries to type something, and hits return. But the Anarchy system is not responding to commands. Who knows why?

PLIK

Eli is pounding the keyboard, but nothing happens. It's a disaster! He breaks the connection, calls back again. This time the Anarchy system is not responding at all, it's acting just plain dead. Who knows why. A disaster.

Paul says, "Damn. It crashed."

Nobody ever heard from The Graduate again. He never posted another message on a bulletin board, at least not a message that any of the boys in Eli's room had heard of. About a week or so after Anarchy went down, Paul tried calling the board's phone number, and it was disconnected.

"The number you are trying to reach is no longer in service." The phone company Kaddish.

Paul wondered if The Graduate ever told his parents what happened, or if he waited in fear for the Secret Service to return. The other boys didn't mention the incident, didn't talk about how whatever they did that day got away from them. It was an odd experience, crashing something. Even a lamer board. Never would any of the three have harmed a system intentionally. Never would any of them have violated the hacker ethic by destroying anything. They only mentioned the experience obliquely after that. Whenever something struck them as weird and inexplicable, this is what they would say: "Plik."

The summer had turned out to be a great one. By the time August rolled around, Paul, Mark, and Eli felt as if they'd known one another forever. They'd learned a lot, and there was more to learn one night in August, when New York Telephone's employees walked off the job to form strike lines.

At midnight, New York Telephone management changed the network password to connect to COSMOS to protect the system from disgruntled workers. Now, in the middle of a session, Paul, Eli, and Mark were suddenly locked out.

How dare they!

At 12:20 A.M., the phone company's Technical Assistance Center got a call.

"Hi, I'm a craft worker and I'm trying to finish this job, see. And you changed the password on me."

The manager who answered the phone had to admire the worker's dedication. We're not going to let a little thing like a strike get in the way of customer service! "Stay on the job, stay on the job," the thankful manager said. "Here's the password—Y6NEQ2."

By 12:21 A.M., the boys were back in business. They were on a roll, they couldn't be stopped. They were all-powerful. Where were they headed? Where would it end? It didn't matter, because they were going together.

FOUR

Plik indeed. Now, if Tom Kaiser could have seen the kind of activity that the boys in Eli's room were engaged in during the summer of 1989, the events that were to follow, that changed all of their lives, might never have occurred.

But Kaiser couldn't see through walls.

As a lawman on the electronic frontier, Kaiser did have some unusual powers, to be sure. He could attach one of those black boxes, a Hekemian Dial Number Recorder (DNR), to your phone. He could keep track of every call you made, every number you dialed, every time you lifted the receiver off the hook.

But Tom Kaiser was not a magician. The security specialist for New York Telephone had no way of knowing the identities of the trespassers he tracked through the labyrinthine confines of the phone company's privately owned, privately operated computers. To him, all the criminals who broke into his computers were equal. All he could do was follow their footprints. And all the footprints were the same size.

Kaiser was not a policeman in the traditional sense. He didn't carry a gun or handcuffs, nor did he have the power to arrest anyone. But he was responsible for maintaining order across all the millions of miles of phone lines that New York Telephone owns. He patrolled the busiest beat in America, because it was his job to keep the peace on that last mile where twisted copper wires connect every home and business in New York to the rest of the world. If you broke into New York Telephone's system, it was Tom Kaiser's job to track you. And stop you.

And he would.

On this morning in 1989, Kaiser arrived early at his twenty-third-floor office in New York Telephone's headquarters building, smack in the center of one of midtown Manhattan's most spectacular intersections, Forty-second Street and Sixth Avenue. He glances out the window at the river of traffic that cuts north up toward Central Park, and then logs on to his computer. Kaiser doesn't waste time. He wants to see what his hacker was up to last night.

Columns of numbers fill Kaiser's screen, a record of all the phone calls that a certain young man in the Bronx known as The Technician made during the past twelve hours. Many of the numbers are familiar to Kaiser by now and are harmless. But one of his chores is to chase down the numbers he doesn't recognize. This morning, he sees one.

By its prefix, he knows it's an internal New York Telephone number. Quickly, he punches it into another database, which lists every assigned telephone number in the New York region. The screen flashes:

```
NEW YORK TELEPHONE CO DIAL HUB
```

The words *dial hub* mean nothing to him. He calls the number, and a modem answers. This is definitely not good.

Luckily, the one person in the world who can tell Kaiser how bad the situation is has just arrived at his office. Kaiser calls him.

"Hey, Fred," Kaiser says to his partner. The hacker has hit a new number. He reads it off, then asks, "What's a dial hub?"

Fred Staples is incredulous. "They were in our dial hub?"

"Yeah, what's a dial hub?"

"Aw shit," says Staples.

"What?"

"How long was the call?" asks Staples.

"About a half-hour."
"Aw *shit*."

For Tom Kaiser, the whole mess had started with an anonymous letter he got by way of AT&T.

Somebody had sent the long-distance carrier a single sheet of computer printout paper. Just a couple of paragraphs, no date.

The letter was sent a few months ago, back in November of 1988. The letter said some kid in the Bronx—The Technician—was getting himself into trouble, hacking into telco switches and God knows what else. Kaiser got a lot of anonymous letters. Usually these turned out to be from some landlord who wanted to use a pending phone company investigation as an excuse to evict a tenant. This letter, though, had been sent by someone who clearly cared for the kid in the Bronx. It had a stop-him-before-he-hurts-himself tone. (Kaiser would always suspect it had been written by a relative.)

The first thing he'd done when he got the letter, the first thing he always did, was to pull up the subject's billing records. And he got a surprise.

The kid was calling AT&T switches, from his home, making no attempt to disguise his trespass.

Up went the DNR on the kid's phone line. The black box was no bigger than an old-fashioned adding machine. It spewed out old-fashioned paper tape, too, making a reassuring clack-clacking sound all day and all night long. An old-fashioned machine to catch new-fangled criminals. Over the years, Kaiser had become familiar with the noise; he could hear it from across the hall, and whenever he was on a particularly hot case—a hostage standoff, say—he would bolt from his seat at the sound of it. It is the same kind of device police agencies use in drug investigations. Unlike the police, however, phone companies don't need a court order to install a DNR on someone's line. That's because the phone company owns the line, and by federal statute has the right to monitor its property.

AT&T's problem with the Bronx hacker had become New York Telephone's problem in early 1989, because one morning when Kaiser was reviewing the night's activity, he found that the Bronx kid had called New York Telephone computers.

Now, this was something new. Kaiser knew all about toll fraud, of course, because for years he had been tracking people who stole phone calling card numbers. But in 1989, it was a relatively infrequent phenomenon to have people breaking into the phone company's computers. Kaiser had never before needed to become an expert in the internal workings of the phone company's own system. So he needed someone who was an expert.

Coincidence saved him. (Kaiser is the first to point out that coincidence has long played a serendipitous role in his career.) The day Kaiser needed to know about New York Telephone's own system, it so happened that an expert named Fred Staples was in the office. Staples was replacing the old clack-clacking DNRs with new equipment—computer DNRs to track computer criminals. Nobody knew more about the phone company's computer security than Staples.

They didn't know it then, but with this hacker case, Kaiser and Staples were going to be in each other's faces for the next few months, like a middle-aged married couple—though they looked more like the Odd Couple. Kaiser is tall and thin and silver-haired, and Staples is shorter, dark-haired, and built like a pit bull. Kaiser's eyes smile and welcome you, reflecting his early years spent in customer service. Staples's eyes bore into you and analyze.

Staples was an engineer and spent all his time building hardware and software for the phone company. He was New York Telephone's main defense against hackers, and it was his job to make sure that what he built was secure. If an unauthorized user broke in to the system, Staples took it personally. It was *his* system.

Kaiser and Staples soon noticed an unusual pattern in the phone calls The Technician made from his Bronx home. First, the hacker would call a phone company computer, stay connected for ten or fifteen minutes, then log out. Next, he would call a certain phone number in Queens. Then he would call back the computer for ten or fifteen minutes. Then the number in Queens. Then the computer. Then Queens.

They pulled the billing records on the Queens number and learned that it was assigned to the residence of Charles Abene.

Kaiser and Staples concluded that someone at the Abene residence must be coaching The Technician. Whenever the Bronx hacker got stuck in the phone company computer, he'd log out and call the Abene residence for assistance, then try the computer again. All night long.

Kaiser got the chill.

He was familiar with the part of Queens where the Abenes live, a neighborhood near Roosevelt Avenue in Jackson Heights. Law enforcement agencies often presented the phone company with court orders for DNRs and wiretaps on the phones of suspected drug dealers in that neighborhood. Kaiser had no idea who was on the phone at the Abene residence. He had no idea it was a skinny kid with a dog-eared copy of *The Hobbit* on his bookshelf. For all Kaiser knew, it could have been some narco terrorists from the Cali cartel who had the run of New York Telephone's internal phone system. They could knock out phone service to the entire Northeast if they knew what they were doing. And imagine what they could do if they didn't know what they were doing.

So now The Technician had gotten into a dial hub. Probably with the help of his friend, Abene.

A dial hub is supposed to be one of the most secure entry points into the whole New York Telephone computer system. In 1989, it was a brand-new technology that the phone company was installing, so new that its name, at least, was unknown to most of the people who worked for the phone company. The hub was a way to keep intruders out of the system even as authorized employees could connect from home. It was a first line of defense and you could only get in if you knew the right ten-digit password. Every single one of New York Telephone's thousands of employees had a personalized ten-digit number.

Once you got into the dial hub it became, simply, a subway system that you could ride to any destination within the New York Telephone Company system. Technically, the hub was a pool of modems. The modems were like trains, waiting for passengers to board them. Next stop MIZAR. All aboard for PREMLAC. COSMOS, last stop. If someone without a token tried to board, the dial hub would disconnect the unauthorized user.

But during the night, The Technician had gotten hold of a token. And for half an hour, who knew where he'd ridden? At each stop, you needed another password to enter a specific system. But the DNR had no way of tracking a rider once he passed through the dial hub's turnstile.

Kaiser asked Staples what the hacker could have been looking for. Staples didn't want to consider the possibilities, but he had to.

"I don't think there's any computer worth mentioning that we own that's not accessible through there," Staples said.

What were some of the stops on the subway? Computers that were used to provision new services. Computers that were used to maintain phone company buildings and plants. Computers that detailed work schedules for every employee and every job. And that was just the administrative stuff. Beyond that, the hackers could enter every switch in the New York Telephone region.

Staples hung up the phone in his office two blocks north, where his window faced Kaiser's. They could have waved to each other if they'd thought of it. They never had.

Staples came right over to Kaiser's office and within an hour he and Kaiser met with Kaiser's boss. They laid out the whole thing: We think we may have something big here. It's something bigger and broader than anything we've encountered before.

This is the moment when the case transformed itself. The day before it had been a case like others in the past, a case of one intruder, a case where only one switch—one isolated point in the network— was hit. Today, at least two interlopers were involved, and they could be anywhere in the system. The number of perpetrators had doubled overnight, causing the case to grow geometrically. Today, the lawmen were fighting a network of intruders. Today, they were facing a conspiracy.

Kaiser's boss agreed. It was time to broaden the investigation.

Up went another DNR, this time on Abene's phone.

Staples told Kaiser how to watch over the hackers' shoulders even after the DNR lost track of them in the system. Staples told Kaiser which administrators were responsible for monitoring the breached computers, and then how to notify the administrators each time the hackers log in. The administrators often could reconstruct, from their audit logs, all the commands that the hackers used to move around in the system. That way, if the worst should happen, someone would be able to work backward to fix the problem.

When New York Telephone's union employees went on strike in August, Staples's strike assignment was to move down to the desk

next to Kaiser's. It was close quarters in Kaiser's office, which except
for the spectacular view looked perfectly ordinary with its Max
Headroom coffee mug and standard-issue bookshelves spilling arcane
technical manuals, just like any other sugar cube of a workspace in
midtown Manhattan. With all the high-tech activity, you'd expect to
see something like the war room of the National Security Council.
You'd certainly expect to see something with more bells and whistles
than Kaiser's ancient black computer terminal, so clunky it could
have been manufactured in Russia. The walls of the office were dirty
white, or maybe clean gray. Two battered metal desks sat in a space
so tight that their edges almost touched—a space that, come to think
of it, wasn't any bigger than Mark Abene's bedroom. But it was the
antithesis of Mark's bedroom.

Kaiser and Staples got along well. Both were life-long members of
the phone company family, workers who got their first entry-level job
with New York Telephone in the late 1960s, back when the phone
company was expanding every day and career opportunities seemed
limitless. Back then, AT&T owned the whole phone system, including
New York Telephone, and it was easy to get a job.

Kaiser was one of the first males hired to handle customer
complaints in AT&T's downtown office at 195 Broadway. He worked
his way up to management, and in the 1980s was transferred first to the
regulatory department, which dealt with companywide complaints, and
then to the toll-fraud department.

As a young man, Staples had worked as a stagehand, fitting together
pieces of scenery like so many pieces of a puzzle. He'd always liked to
build, had always seen the world as being divided into units that could
be assembled. The phone company hired him as a communications
serviceman in the late 1960s, when rapid expansion had overwhelmed
AT&T's mechanical capabilities. Customers complained about bad
service. Staples learned to repair teletype machines. As the phone
system changed, and grew, so did Staples.

Around the same time that the boys in Eli's room were crashing the
Anarchy system, the DNRs were leading Kaiser and Staples to two
more phone lines. By now, the pattern was familiar. By now, it was not
such a shock to learn that there was a whole group of trespassers out
there. By now, the lawmen had gotten used to a word: *conspiracy*.

The two new phone numbers were ones that Abene dialed frequently, repeatedly, in between calls to phone company computers. So now the black box watched the lines of Elias Ladopoulos and Mrs. Jean Stira as well. You wouldn't believe the stuff that was turning up on the morning logs. Dial hubs were the least of it. The Technician had turned into small fry, bait to catch bigger fish. Kaiser and Staples feared they might be dealing with a widespread threat to their network's security, that there was nowhere the hackers couldn't—and wouldn't—go.

During the strike, Kaiser and Staples worked eighteen hours a day, seven days a week on the hacker case. Staples sat by the window, chain-smoking Marlboros, trying to get inside the hackers' heads.

The situation was quite impossible. The hackers were probably kids, just like The Technician. But maybe not, maybe they were drug dealers. Should Kaiser and Staples shut down the hackers now, just block their access to the phone computers? Of course, if they did that, they wouldn't know how far the intruders had gotten, nor what they were after. But if Kaiser and Staples allowed the hackers to continue, couldn't some awful consequence result? What if the hackers crashed a system? It was kind of like having a tarantula crawl up your leg. If you shook it off too fast, it would escape into the wall. But if you waited too long, you got bitten.

They wanted help from law enforcement. They wanted to send a message that hacking the phone system was not the touchy-feely thing it was back in the 1960s when Joe Hacker would get a friendly "wakeup" call from the phone company, a polite warning. Too much was at stake now. You'd hit the wrong key, and oops, there went phone service to Wall Street.

Kaiser and Staples knew that unless they could show that the intrusions had caused at least a thousand dollars' worth of damage or loss of service, they wouldn't get the U.S. Attorney's office interested. And it was important to prosecute. If the hackers were kids, it was still a good idea to send a message, scare them, come down on them with the full, heavy weight of the law. Kids wouldn't end up in jail for this kind of bravado trespassing, they figured—hell, they had kids of their own. But the hackers would learn their limits if they got a stern lecture from a judge, coupled with a couple hundred hours of community service.

Since 1987, the phone company had, in fact, snared other teenagers snooping around in the phone system. Kids with names like Bill from RNOC and Delta Master and Ninja NYC. Those cases, involving juveniles under the age of eighteen, were disposed of quietly. Sometimes with only a phone call to a parent.

That's what happens if you're dealing with kids. That is, if you know for sure that you're dealing with kids.

Staples spouts a blue stream of Marlboro smoke, considering the options. With Staples in the office the smoke gets so thick it reaches down to the floor. The room was in a haze. Months from now, Kaiser would move a picture on the wall and see the darkened outline of the frame's original position. Staples smokes so much that it gets into the paint.

One option is going to the media, going public with the fact that it is nearly impossible to fool around undetected in phone company computers. Announce the intrusions and at the same time show how the system has been secured. Another possibility, assuming the hackers were all kids, is creating an educational program to channel some of this misdirected energy. The phone company, after all, has a hard enough time finding qualified people to train as telecommunications engineers.

But that's academic. And this is the real world. The facts for Kaiser and Staples are these: We don't know who we are dealing with. And we are witnessing a more dangerous level of infiltration than ever before.

In the end, they realize there is only one option.

Kaiser awakens early one morning and thinks, today could be the day that we nab the hackers. He already feels the summer heat through his shirt as he leaves his house on Long Island to make an early train to Manhattan.

This morning he will have to convince more literal-minded lawmen that crimes committed on computers are as much their concern as the drug buys on Forty-second Street.

Today, Kaiser has reserved the big conference room, and the meeting is full of phone company people, and investigators from the

New York City police department's special frauds unit and the U.S. Secret Service.

Knowing that the technical aspect of the case might make it difficult for non-technicians to grasp, Kaiser and Staples explain the case in general terms.

"We may be coming to you with this case we have, and how should we do that?" Kaiser says.

"What have you got?" asks one of the investigators.

"We have three hackers," Kaiser says.

Staples gives it a shot and talks about how the phone company switches are being targeted. He talks about the dial hubs, and about how it would be relatively easy for New York Telephone to simply close the holes in the system to lock out these particular hackers. But the problem is bigger than that, he says. If the hackers could get into New York Telephone's computers, they might be able to get into the phone system in other regions as well.

"The problem is national," Staples says.

The investigators are taking all of this in, nodding, jotting down notes. Somebody says Staples and Kaiser should meet soon with the U.S. Attorney's office. A Secret Service agent says he will report this to his supervisors.

Staples and Kaiser are feeling pretty good, like they got the message across and everyone in the room understands the magnitude of the problem.

Then one of the investigators raises his hand. All of the New York Telephone Company people look at him.

"What's a switch?" he asks.

It was autumn now, and the heat of the city summer seemed to have been absorbed by the people walking the picket line that Kaiser has to cross to reach the revolving glass doors of his skyscraper. Crossing the picket line really doesn't bother Kaiser. Sometimes they threw eggs, but in his teens Kaiser was a Teamster. He could handle a few eggs from the Communication Workers of America.

One day in October, late in the afternoon, the black box told Kaiser that someone from the Ladopoulos residence had just placed a call to New York Telephone's business office. Right under his very nose. And it was still in progress.

What were the hackers after now?

It was a long phone call. Kaiser thought the connection would last forever, the minutes were just ticking by, and he was dying to find out who the hackers were calling at his shop.

Finally, the DNR told him the connection was broken.

Immediately, Kaiser punched the digits to connect him to the same number at the business office.

The employee who answered the phone was an old supervisor of Kaiser's, from the days when he worked in the business office. Normally, she wouldn't be answering the phone at all, not when Kaiser calls and not when a hacker phones. But her strike duty was to take field calls that came to the business office.

"Who was that on the phone?" Kaiser asked her.

"Some poor plant guy who's stuck up on the pole," she said. She was surprised that Kaiser was calling and that there was a problem with what seemed like such a routine request.

Kaiser winced as she explained that the caller said he was a repair technician named John Gilmore who needed a phone number transferred from New York to New Jersey. Of course she'd put through the work order. The request was perfectly ordinary. Kaiser recognized the name John Gilmore: Gilmore was a former hacker who went on to become a millionaire writing code at Sun Microsystems.

Kaiser quickly countermanded the work order.

Every day was an exercise in frustration. It seemed like the case would drag on forever, and all Kaiser and Staples could do was run around putting out fires. By the end of the summer, they thought they had accumulated enough evidence, and they finally met with an assistant U.S. Attorney right after Labor Day. They went downtown to his office, scrupulously prepared for the meeting, organizing all the information they had collected about unauthorized intrusions in notebooks with neat colored tabs. They had charts, too.

The prosecutor told them his office was definitely interested. Just keep accumulating evidence.

More meetings followed—uptown, downtown, in Brooklyn, where the Eastern District of the U.S. Attorney's office is headquartered.

Meetings with the Secret Service near Wall Street. Meetings with the police.

The tarantula was still creeping.

Just sitting there, forced to watch the hackers logging in to their computers, was starting to drive Kaiser and Staples a little crazy.

They decided to go undercover in cyberspace. The only disguises they needed were anonymous computer handles. They had some phone numbers for so-called underground bulletin boards, so they called one named Shadoworld. They logged right in, just like hackers would.

They used the handles Splinter and Rapier. They felt a little silly doing it, being grown men and all, but at the same time, it was useful. They cruised the philes, and got a real sense of the kind of information that your run-of-the-mill wannabe hacker possessed. Kaiser said, "It increases our level of knowledge."

They feel like spies, reading philes and looking for clues that the hacker intrusions might be more widespread than anyone suspects. Who's to say this case is limited to just three hackers? Who's to say it's limited to phone company computers in the New York area?

Evidence of a broader problem might interest the U.S. Attorney's office, they figure.

Then, one day they are in the prosecutor's office, explaining the case again. Some of the faces are old. Some of the faces are new. It seems clear to everyone that the three hackers have repeatedly entered phone company computers. It seems clear that the evidence is irrefutable. Kaiser and Staples are ready for a sign, ready for search warrants, ready to put an end to the damn thing.

And someone raises a hand, and asks, "What's a switch?"

FIVE

Sometimes you just have to blurt out the news, not even try to cushion the impact. When the whole world shifts and heaves, you just have to come out and say what happened.

Mark got kicked out of the Legion of Doom.

There it is, in all its horrible bluntness. He's *kicked out.* He's gone. He's unplugged.

Word of his expulsion leaks out, all over the underground. How did it happen? Some kind of a fight. Who knows? But it's posted on bulletin boards from here to Germany. It's the talk of the hacker elite: Phiber Optik got into a feud with Erik Bloodaxe, and to hear Erik Bloodaxe tell it, Phiber Optik lost. Here's how it happened.

One day in 1989, while Chris is working on his big hacker project, a directory of the computers on a large data network known as Telenet, the phone rings.

The caller is LOD member Mark Abene, up in New York City.

Mark is really upset. His account on the NYNEX Packet Switched Network was killed. Can you imagine? Phiber Optik without access to the NYNEX Packet Switched Network. It was like James Dean without a motorcycle. Mark desperately wants to get back in the system, and knows that Chris has a secret route to the computer. Mark asks for it.

Now, Chris knows Mark has access to a list of addresses of certain phone company computers that you can reach over Telenet. Chris wants to include those addresses in his directory.

"Why don't you just go on there and get that and give me the list?" Chris asks Mark. Chris is calling the shots, after all.

"I can't do that right now, but you can just go do it yourself," Mark says, giving Chris an account name to get the list.

When Chris tries, the account doesn't work.

"All right, I'll try to find another one," Mark says.

But now for the outrageous part, the part that will make Chris sputter with rage even years later.

After they hang up, Mark calls Bob, a hacker friend of Chris's who lives in Massachusetts. And Mark asks Bob for the secret route. "Chris said it's OK," Mark convinces him.

Bob calls Chris later that day, and says, "Well, Mark called up and I gave him the information."

"You did *what?*"

"He said you said it was all right. He was really convincing."

"Did he give you the account?" Chris asks.

"No, he said he was going to give it to you later."

Chris is totally furious. Right away, he gets Mark on the phone. "Mark, what are you doing calling Bob and telling lies to get that route? Give me the account like you promised. I *need* that list."

"I don't owe you *shit!*"

There it is again, that New York attitude.

"Excuse me?"

"I don't owe you shit. I didn't get it from you. I got it from Bob. Fuck you."

"*What?*" Chris says.

"No, man, I don't have time for this," Mark says, hanging up the phone on *Erik Bloodaxe!*

Well, you can imagine how mad Chris was.

He calls every single member of the Legion of Doom and tells them what went down. One by one, he gets their votes.

Mark is out.

Out.

Out.

"You don't screw people like that," Chris says.

Of course, Mark told his own version. His version rambled a little.

Let's see, from Mark's perspective, the name Legion of Doom had outlived its reputation. He was so far ahead of what anyone in that gang knew that it was ridiculous.

LOD had its own bulletin boards, of course, which only members could visit. Catch-22, the first LOD board that Mark logged on to, was just one of many. You also got access to Plovernet. But plenty of good boards weren't controlled by LOD. Sherwood Forest, which actually was located in Forest Hills, Queens, was run by a kid who called himself Magnetic Surfer.

Mark told people he got bored with LOD. In his opinion, the text philes on its boards were moronic, supposedly instructing you in how to hack COSMOS, how to hack a Unix computer. Utter silliness, in Mark's opinion, written by pretenders who knew so little about their subjects that they listed incorrect commands, wrong descriptions.

Who wanted to be associated with that? Did Mark actually say he quit?

Well, in Mark's version of the story, yes, there was this guy, and yes, his name was Chris Goggans, and he ran around acting like he was in charge of LOD. Like he was the boss. Everybody up north had heard he was some rich kid who lived in the Lonestar State. Of course no one really *knew* this because no one had met him face-to-face. All contact was through bulletin boards and phone calls.

Mark heard that Chris was talking Mark down. He was furious with Mark, for some reason, and Mark heard that Chris was going around posting stuff on bulletin boards, telling people not to trust Mark with information. Chris was saying that Mark *was not in the LOD!*

Chris was telling people that Mark cheated him out of information.

"Not true," Mark said.

Mark doesn't bother to argue much about what Chris said. What does he care? he tells people. Still, it has to hurt. He has a personal phone number: 949-4LOD. What was he going to do, change it to 949-XLOD? It's pretty clear that Chris had been out to get him from the beginning. And it wasn't fair. That has to hurt.

Mark had other problems. He'd just changed high schools again. Don't ask. It was a long story, about how the first public high school, Francis Lewis High School, cracked down on him for breaking into the computer where they keep the grades and attendance records. He didn't change anything—the scientist was just making a few observations—but his mom had to come in for a meeting, and Mark got suspended. They were pretty rigid at that school.

Then the second school, Newtown, didn't give anybody a locker, and Mark had to carry all his books around all day long, in a big black bag with the logo "UNIQUE" written across it in rainbow colors. Newtown scheduled Mark's math class at 7:30 A.M. It was kind of hard to learn trigonometry when you felt like puking, and Mark was queasy in the morning. He spent most days cutting school, smoking Benson & Hedges Ultra Lite Menthol 100s (the really long ones), and eating buttered bagels in the pizzeria across the street.

The pizzeria was a walk-in place, no chairs, so Mark and his non-hacker friends Howie and Gustavo and Jaime, who lived up the block, would sit on the floor and philosophize. They argued about the speed of light. When the pizza place eventually burned down, the principal came out on the street, rejoicing.

The counselors at Newtown suggested that Mark might do better in an alternative school. His aunt had heard of City-As-School in Manhattan. This is a place that doesn't resemble any school you ever attended. It's one of those places you have to come to New York to see. No classes, no strict schedule, no patrols in the hall. To graduate, all Mark had to do was complete some practical internships around the city. You see, the city *is* the school, get it? He had to keep a journal, the worst part. And he had to check in with his adviser once a week. The rest of the time, he was an intern. He Xeroxed papers at the city's Department of Investigations, or worked at the Queens Hall of Science, explaining the inside of the eyeball to groups of kids on field trips. It was his favorite exhibit.

There was another exhibit at the museum, one Mark really liked. The Anti-Gravity mirror, and if he stood perpendicular to it, it looked like he was floating.

* * *

But if Mark could float away from the Legion of Doom without a look back, Eli couldn't. The feud pissed Eli off, frankly, on behalf of his friend. Who did these LOD guys think they were, dissing Phiber Optik of the LOD? Why did everyone say the Legion of Doom was so great anyway? What made them special?

Eli had had a lot of friends in LOD, the original LOD, back before Goggans took control. Now it seemed like everybody was in a gang, and most of them were the saddest excuse for hackers that Eli had ever seen.

Maybe the Legion of Doom was elite simply because it existed. If it was not the only gang out there in the underground, it was certainly the only one that boasted a decade of history and dozens of members nationwide who answered to a Texas leader. It *was* the biggest. It was the baddest (that Eli knew of so far). But if another group of hackers challenged the Legion's superiority, well, who knows what might happen.

Eli had been thinking.

There was an idea he's had for a while, it turned out, and one day he told Paul on the phone.

"MOD," Eli said.

"Mod?" Paul asked.

"M-O-D," Eli spelled.

So what was it?

"We should call ourselves MOD," Eli said. It was like a joke, a finger in the eye of LOD. He explained that it was an allusion to LOD, the Legion of Doom. From L to M, the next iteration, the new "kewl dewds" of cyberspace. The boys from New York were the *opposite* of the boys from Texas. How better to define themselves? The boys from New York could figure out who they were simply by opposing the Legion of Doom and everything it stood for. Whatever it stood for. That wasn't important right now. What was important, Eli said, is the joke: MOD. It stands for nothing. It stands for everything. Masters of Disaster. Mothers on Drugs.

Masters of Deception.

Paul thought it was a great joke, a great way of spoofing the Legion of Doom's cachet. Everyone had to belong to something.

Later, Eli started writing a phile called "The History of MOD."

He wrote:

```
MoDmOdMoDmOdMoDmOdMoDmOdMoDmOdMoD
^^^^^^^^^^^^^^^^^^^^^^^^^^^^^^^^^^^^
        [The History of MOD]

      BOOK ONE: The Originals
```

MOD: it was a mysterious name and it needed some good narrative to fill it out. But these gang members didn't go anywhere, physically, or do anything, physically. Mainly, they sat around in their bedrooms, talking on the phone and typing on their computers. They also went to the mall.

"The History of MOD" had to elevate the gang to the status of a modern-day Magnificent Seven, riding into the sunset of cyberspace. And why shouldn't it? The MOD boys were definitely riding the ragged regions of the electronic frontier, even as the whole territory was still being carved up. It was the Wild West, right? There weren't any laws yet that would tame these desperados. Information was out there for the taking, and they were grabbing.

How to bring it all to life. This was Eli's quandary. He recounted what happened after he met Paul and Hac:

> Soon the three were partaking in all sorts of mischievous pranks, and . . . they took to knocking down the locals who thought "I know all there is about hacking." It was in the midst of all this fun that they agreed to form an underground group called MOD (approx. June 1989).

But looking for the stuff of legend here, Eli wanted to introduce the group's golden member, formerly of the Legion of Doom, ladies and gentlemen, Phiber Optik.

Eli wrote about how he and Mark became friends. He would always remember a challenge he'd issued—a challenge that Mark so easily crushed. If you're so great, Eli told Mark the first time he phoned him, try figuring out who I really am. Here's the only clue, my phone number: 555-ACID.

Now, 555-ACID was *not* the home phone number assigned by New York Telephone to the Ladopoulos family. It was a number specially created for Eli on the Hollis switch (and unknown to New York Telephone). When a caller dialed this customized number from anywhere in the world, the call was forwarded to the Ladopoulos's real—and unpublished—number. So a caller could connect with Eli but not know who Eli really was or where he lived.

Armed with 555-ACID, and a phone number to connect to the Hollis switch, Mark went to work. Of course, for Mark, the task was moronically simple.

Within minutes, the phone rang in the Ladopoulos household.

Mark calmly read off the Ladopoulos's real phone number, as well as selected billing information. You had to be impressed.

Eli was:

> Another hacker and telco computer specialist also seemed to be very prominent and knowledgable then as well. He wasn't liked very much, because he seemed to have a rather large ego, which I may add, makes it okay to have when you know so much as he did. He declared he was "Phiber Optik of the LOD!" Scorpion, Acid, and Phiber exchanged ideas on switching theory for a long while, but then came the time when PO wanted to know Acid Phreak's phone number since he found it "unfair." AP mentioned that he could prove himself by finding it for himself. Armed with a dialup, PO called Acid back on his real number and casually proclaimed victory. And so, Phiber Optik was "brought into" the group.

Months later, when Mark would actually see "The History of MOD," he would smile, the way he always did, and pronounce it all "nonsense," the way he always did.

But it grew and grew and grew. It was the nonsense that legends are made of, like Billy the Kid. "The History of MOD" would be widely disseminated, passed along in one perfect copy after another, freely distributed across the electronic bulletin boards of cyberspace, where publishing costs nothing. It was read more hungrily than any dime store western.

Eli loved to write. He always had. He created MOD as a joke, and then the joke became real. He kept adding new members to the joke, until after awhile it wasn't a joke anymore. It was a gang. It attracted some of the quirkiest talent to roam the wired wilderness.

Like Supernigger. Eli met him on a New York City bulletin board called The Toll Center. The handle was not the most politically correct, nor did it accurately invoke the person who uses it. Supernigger—according to some people who claim to have met him—was a slight white kid, a teenager who affected a remarkable array of accents and voices to trick people into giving him information. Nobody was a better social engineer than Supernigger. He held the world's speed record for talking a phone company employee out of his password. Supernigger just called up, said, "This is Bob from Service. What's your password?" He got it, clocked the whole call under ten seconds. Usually he spoke in a lazy southern drawl, telling the woman in the business office, "Lady, I'm twenty feet up on the pole." She gives him whatever he wants. On a BBS, Eli tells Supernigger to give him a call, at (718) 555-ACID. Supernigger dials, connects, and is blown away. Imagine having your own customized phone number. Absolutely free. Phone company doesn't even know it's there. Sign me up.

Supernigger, it turned out, had social engineered the phone number of a conference bridge. A conference bridge is a lot of fun. Big corporations use conference bridges all the time. They're really just big party lines for companies. A dozen people in different cities can be on a conference bridge at once, hearing the chairman of the board predict the quarterly earnings. Supernigger gave the number to everybody in MOD, and the MOD conference bridge was born. Every day, at the same time, all the MOD members would call the number and talk to one another. It was way cool for a while. But after a few days, it got kind of boring. What good was a conference bridge if nobody knew you had one? So they started phoning lamers. They called preadolescent dorks who posted stupid things on BBSes, giving real hackers a bad name. Somebody on the MOD bridge dialed a lamer, and then a dozen kids all started screaming into the phone, rampaging, scaring the kid. And if you were really lucky, his mom would get on and say, "David? Who is it?" She'd be all confused, as a dozen voices caterwauled, "David? David? Who is it?" And then someone would yell, "It's the MOD!"

The name started to get around.

Eli met a lot of people. He got friendly with a hacker who used two handles: Thomas Covenant and Sigmund Fraud. But everyone knew him as Pumpkin Pete, a nickname he hated but couldn't shake. He's from Brooklyn, but he joined the Air Force and got stationed in Florida. He was obsessed by the phone system, was so desperate to get phone company technical manuals that he pulls an outlandish stunt. He tries to order hundreds of manuals from Bellcore, the research and development arm of the seven regional telephone companies. He tells Bellcore that they're building a new central office in Florida and they need the manuals right away. It doesn't work—it just prompts a military investigation. But the stunt's good enough to give him a reputation. And besides, he's from Brooklyn. Now he was in MOD.

Paul's friend Hac is also inducted into MOD. He was good to have along on trashing expeditions.

Eli and Mark spent a lot of time cruising bulletin boards. They rode out across the vast wilderness of cyberspace, staking claim to anything they discovered. They came to think of it all as their own territory. They defined themselves as being somehow different, and therefore better, than the other hackers they encountered. They roust lamers and ask competent hands to ride with them. Mark thinks of himself as "maintaining" the area.

When the MOD boys come across any new bulletin board that aspires to be a hacker's board, they phone the system operator to give him the once-over, to see if he should be shut down or invited to join. That's how Mark met a kid named Zod, who was running an inane bulletin board in the 212 area code. Zod is a great name, borrowed from the evil genius who attempted to thwart Superman (in the movie). Zod's got his own gang, called Ace, which Eli is scouting for local talent.

Mark logs in to Zod's board as Phiber Optik. That should be enough of a calling card. Everyone in the underground knows who Phiber Optik is by now.

But Zod disses Mark. No one automatically gets access to this board. Not even Phiber Optik. Zod demands that all aspirants fill out a *questionnaire* first to show they are worthy. If you don't know the answers, you don't get a password. Zod has larded the questionnaire with phone company acronyms. What does PREMIS stand for? What does MIZAR stand for? LMOS?

As far as Mark could tell, Zod didn't know the answers himself.

Mark, of course, knows what they all stand for, as well as what they do. Angrily, he logs out of Zod's board.

He takes a tour of those same acronyms, in fact, to figure out who this Zod character really is. He logs in to one system, where he finds out the cable and pair number for Zod's bulletin board's phone line. With the cable and pair number, Mark checks into another phone company computer, where he sees another phone number assigned to the same house. He still doesn't know the name of the subscriber, but he dials the number anyway.

A man answers the phone, and believe it or not, instead of saying hello like a normal dad, he gives his whole name. First and last name. His last name's Perlman. Mark asks for Mr. Perlman's son.

Zod gets on the phone.

"Perlman," says scarily deep-voiced Mark. "This is Phiber Optik."

Zod is flabbergasted. "How did you get this number?"

Mark says, in the understatement of the year, "I looked it up."

One kid who Eli met was really aggressive. He was pretty young, but he'd already taught himself a lot. This kid called himself The Wing, on a bulletin board called Altos. Altos is really one of the main boards for young cyberdudes. It's like Dodge City. Every aspirant, every kid with a modem and a desire to hack hangs out there. The Wing said he was a specialist in Unix, a language spoken by most telephone company computers and many computers on the international web of computer networks known as the Internet.

You needed a Unix guy on the team. Unix is seductive to budding hackers because it's so versatile. A computer running Unix can "multi-task," run many different applications simultaneously. But there's a beauty to the cryptic Unix language that's just as important. Unix is like the King's English. There's an eliteness attached to any machine that runs it. And by extension, any hacker who knows how to speak it.

Unfortunately, Unix was not that accessible to most kids in 1989, because you needed one of the newest, fastest, most expensive home computers on the market to run it. A Commodore 64 was woefully inadequate. A TRS-80 was out of the question. For Unix, you needed at least an IBM-compatible computer with an 80386 chip and four megabytes of RAM. (Today they're practically giving those computers away at gas stations with a fill-up. But back then, the system cost about three thousand dollars.)

Only kids who had more money than the boys from Queens could run Unix. The hacker who Eli met on Altos was one of those kids.

The Wing lived in Pennsylvania, just over the state border from Trenton. His real name was Allen Wilson. It was easy to strike up a friendship on Altos, because the board had a chat system, a place where a few kids at a time can log in and type conversations isochronously, in real time. You see someone's message appear on the screen as soon as it's typed. He sees your response as soon as you type it and hit ENTER.

On Altos, Eli learned that Allen has his own Unix machine. Allen ran a bulletin board in Unix, if you could believe it. Allen called his bulletin board the Seventh Dragon. He gave Eli the number. Eli wrote:

BOOK TWO: Creative Mindz

> With the addition of Allen, came a shitload of pranks and loads of fun. He hadn't known much about telephone systems, but one thing he knew was how to make Unixes do nifty things. Of course, he and Scorpion had undertaken the task of taking on some worthwhile projects and providing the group with some healthy side-benefits (which cannot be mentioned or commented on at this particular moment in time).

Of course there were things that couldn't be mentioned. That's because they were still going on. And maybe there were things that couldn't even be explained, at least under the hacker ethic.

The "healthy side-benefits" fell under both those categories.

The "worthwhile project" that Paul and Allen had undertaken was this: they were invading a private computer and programming it to find long-distance calling card numbers. It seemed victimless to them. They needed the numbers to fund calls to further their education. Who was being hurt? Not the person whose calling card number got used, because that person would dispute the bill and never have to pay. Not the phone company, because the filched phone calls emanated from a reservoir of limitless capacity. It was like riding the rails. The trains were running anyway, and a hobo wouldn't displace any cargo in the boxcar.

This was how the hack worked. Allen procured the phone number for a computer in the backroom of the Eye Center, which stores and updates customer records. And if you must know, it's where he gets his glasses. Paul doesn't ask Allen how he got the number.

But Paul calls it. Working at his own computer one night long after the Eye Center employees have locked the store and gone home to bed, Paul dials and gets inside this machine that he's never seen. Allen dials and gets inside the machine, too. They're not logged in at the same time, but in a sense they are, since they tell each other what they see. They compare notes.

The Eye Center's machine is just a generic computer connected by a modem to the phone lines. But it has its uses.

Paul writes a program that tells the computer to start working for him every night. What Paul wants the computer to do is what his own Commie 64 used to do for him, scan out useful numbers. That way he can keep the Commie 64 free for other hacking activities while the Eye Center computer works to find International Tele- phone & Telegraph calling card numbers. He tells the computer to dial a seven-digit number that every ITT customer uses to initiate a long-distance call. Then the computer dials thirteen digits that adhere to a known formula ITT uses to generate its calling card numbers.

Usually, the number is invalid. But the computer doesn't care. It hangs up and tries again. All night long. Every night. Whenever the computer hits on a number that works, it records that number. Paul has taught it to keep a list for him.

Early in the morning, say at 5:30 A.M., Paul instructs the Eye Center computer to reboot, that is, shut itself down and turn itself on again. Rebooting automatically erases all evidence of its moonlighting.

Paul and Allen needed calling card numbers because they did a lot of long-distance dialing. To each other, for starters. A call from Queens to Pennsylvania is a long-distance call, and who can afford that twice a night? They called out-of-state bulletin boards as well.

This went on for months, through the summer and fall of 1989. Paul and Allen shared their "side-benefits" with Eli and other friends. In all, the Eye Center computer found about 150 valid calling card numbers. That was a lot of free long-distance calls you could make. That was thousands of dollars' worth of service that ITT wouldn't collect a penny for. The boys in MOD didn't stop to think about it, really, but there were people who might say this was more than just a prank.

The Eye Center hack got so sophisticated that the boys rigged Allen's computer to dial Eli's beeper with each new ITT number. Eli could be walking around and his beeper would go off and he'd look down and—thank you!—manna from Heaven.

There were people who might say things were escalating.

But to the boys in MOD, it was just another good hack. Oh, like it was their fault that the Eye Center didn't know how to make the computer secure? No way. They were just doing what they knew how to do. It was all about information. And information should be free. Shouldn't it?

Of course, when you crossed that line, claiming for your own purposes that it's okay to blame the victim for not protecting himself against the thief, you might eventually find yourself living in a world where you become the prey.

That's what would happen to Paul in the late summer of 1989, although he couldn't know that yet. Paul would become the victim of somebody's "prank," and the consequences would come back to haunt him a few years later.

Some hacker, probably a friend who Paul trusts, gets on to an internal AT&T network of 140 Unix computers nationwide. The network handles administrative chores. The intruder leaves a trail, muddy footprints that will lead right back to Paul and make people think Paul was the one inside the AT&T network. Here's how: First, the intruder adds passwords on a New Jersey computer that's part of the network. The passwords will grant two new users the highest level of access. Those users are named "The Wing" and "Scorpion." The intruder also loads a computer program known as a "logic bomb." The logic bomb keeps an eye on the network's internal clock, and when it reaches a specified time and day . . . plik. The entire system will be wiped out. All that will remain will be this greeting, a greeting that will make people think the logic bomb was set by Paul:

```
Your system has been crashed by MOD
The Masters of Disaster
```

```
Virus installed by Scorpion
GOODNIGHT SUCKERS
```

Now, it was a good thing that one of AT&T's technicians found the logic bomb almost immediately and disabled it. Had the network crashed, it would have been a disaster indeed for AT&T. And a disaster for Paul.

Why would someone sign Paul's name to this piece of work? Anyone who knew Paul would know that malicious destruction of a computer wasn't the sort of thing he would do, let alone condone someone else's doing. But try to prove that. His name is signed to it. This is the sort of thing that happens when you throw your lot in with a bunch of desperadoes. This is what happens when a joke becomes real.

Who among the ranks of MOD would do such a thing? Paul, even after learning of the existence of the bomb, will never know for sure. There are a lot of people in MOD by now, a lot of people who know Paul's handle is Scorpion.

SIX

When you look back at it, there comes a time in any good history where the plot twists unexpectedly. Life is moving along, developing its own routine and rhythm, when all of a sudden, something—or someone—intervenes and the speed accelerates, faster, and faster still. And you hurtle off in a new direction.

One day, while Eli is maintaining the territory, he decides to check out reports he's heard about some hot hacker who lives in Brooklyn.

The Brooklyn hacker calls himself Corrupt. He's rumored to be a specialist. MOD can always use another specialist, and Corrupt supposedly knows more about a ubiquitous and powerful corporate computer called VAX than the founder of the Digital Equipment Corporation. Which would be some feat, considering that DEC *manufactures* the damn machine. VAX stands for Virtual Address Extension. You see, in microprocessor design, there's this concept known as virtual memory addressing, which allows a computer to behave as though it has access to more memory than actually exists. It does this by page swapping, or grouping an arbitrary number of consecutive bytes together to call them a page. Then the microprocessor "swaps" pages, and is able to time-share. It's pretty complicated, but Corrupt instinctively understands the process.

An expert who understands the intricacies and nuances of running

VAX computers could really widen MOD's power base. A VAX master could help the other MOD boys navigate through thousands of computers that for now seem tantalizingly obscure. Not only is VAX a type of computer prized by hackers who love the versatility and power of the machine, but VAX is also indispensible to universities, corporations, small companies, database archives, and libraries all over the country. Oh yeah—the government owns a lot of VAXes, too. The government keeps a lot of its secrets hidden on VAXes.

And Corrupt can crack VAXes. Sign him up.

Now, there was plenty that Eli didn't know about Corrupt. He didn't know, for instance, that Corrupt's name is John Lee. He didn't know that John lives with his mom in a third-floor walkup apartment in Bedford-Stuyvesant (that's *Bed-Stuy,* you've heard of it as surely as you've heard of Cabrini Green and East L.A.), one of New York's toughest neighborhoods. Eli didn't know that John would need no introduction whatsoever to the concept of MOD, because John was intimately acquainted with gangs. Out in the real world, out on the streets where you measure distance with your feet instead of your modem, John used to belong to a gang. He was a Decept, a member of the best-known gang in the whole city. Anytime there was a holdup on a subway train, you can bet some cop was going to say the Decepticons pulled the job.

John got out of that scene. But he knew the value of belonging to a group. He knew the kind of protection that's available from friends. He knew what any Decept learns, that it can be cold and lonely out there in the dark. And there were a lot of dark spaces in cyberspace.

The phone rings.

"Hello," John answers.

"Hey, I hear you know a lot about VAXes," Eli says. No introduction. No "my-name-is" crap. Just a verbal challenge.

"How did you get my number?" John asks.

"I got it from Sage," Eli says.

That's all Eli needs to say. Now John knows he's for real; now John will believe Eli when he introduces himself as Acid Phreak.

Sage is the name of a bulletin board. To get access to the board, John had to tell the system operator his hacker handle and his real phone number. Eli says he's got a lock on the Sage system, he can get into the administrative files and pull up any information that he wants about the board's users. In this case, Eli just dipped into the registration file. And there was John's phone number.

John's impressed, because he's only owned a modem for about six months; he's only been into this hacking thing since then, so yeah, it's pretty cool that Eli has this capability. They talk for a while, about VAXes and other things. They hit it off. Eli says he has a friend John should meet, too. John says that's cool.

The next time Eli calls, his friend Mark is on the line with him.

They all start trading information. They have this phone friendship now, talking back and forth, and they realize they have a lot of information to share.

This is the best thing that's happened to John since he started hacking, frankly. He's met other hackers who are cool, but nobody like these MOD guys.

Would Eli and Mark have been so willing to befriend John if they'd known how *lame* John was a mere six months earlier? Of course, they'd never know, because John was the kind of guy who doesn't let on. He could convince anybody of anything.

John used to be the class clown. He was always the one who could incite people, get them to do things they really wouldn't have considered on their own. Way back at All Saints Elementary School, nearly ten years ago, he used to get the whole Nut Row in trouble. That's when he rode the train every morning to Williamsburg, where black-sooted All Saints Church rose like a thorn from the concrete. It was private school for poor people. A lot of the students were black, like John, and a lot of students lived with only their mothers, like John. The sign out in front says in both English and Spanish: LOW TUITION—CONVENIENT MONTHLY PAYMENTS.

The "Nut Row" was the unofficial name for the segregated group of desks off to one side of the classroom. The bad boys sat in the Nut Row. There was Jimmy Gold in the row's first desk—Jimmy sitting, jiggling, bouncing, the hard heels of his lace shoes drumming against the linoleum of the classroom floor. He couldn't help it. Then, behind him was James, and, you don't want to know, but yes, he had his pants unzipped under the desktop as usual. Behind James, staring blankly at the crucifix at the front of the room, was Ernie. And behind Ernie was the class clown. John was taller and thinner than most of the other ten-year-olds, and he had the kind of ideas that got the rest of the Nut Row into a lot of trouble. Like sneaking off on a private unguided tour of the sacristy after mass.

Every day it was something else. There was the time after the bathroom break when somebody in the Nut Row brought back some feces to his desk. That's a problem. There was the fight in class. That was not good. There was the time in the cafeteria that somebody in the Nut Row swallowed an Oreo, regurgitated it in nearly pristine, original shape, and started a terrible chain reaction of nausea at the table.

How did it happen?

"Hey, James, do the thing with the Oreo," John would say. "Come on, man, do it."

It was easy to get carried away in John's presence.

Was it any surprise that John was asked to leave All Saints before fifth grade? He had so much promise, but he needed more stimulation, the principal told his mother, Larraine. It wasn't that she wanted to get rid of John, Sister Donna Jean Murphy said. She'd seen boys like him before, who just get distracted and go off on tangents.

From fifth grade on, John attended P.S. 11 in Clinton Hill and he didn't think it was so bad. It had a satellite program for advanced students. There was a roomful of computers, Commodore PETs, one of the first personal computers to hit the market, big boxes with glowing screens. (At All Saints, nobody had ever heard of computers.) The classroom at P.S. 11 wasn't as fancy as a room that would be called a computer lab, but John just needed a quiet place to sit and type. Like in Sheila's office.

Sheila was the assistant principal at the high school where John's mom worked, and John had hung out there a lot in the afternoons while he was in elementary school. John liked Sheila's computer. It was an Apple II, and it was the first computer he'd ever seen. How did it work? John wanted to know. The case was a glorious gray-brown molded plastic, embossed with a rainbow apple logo. Today it would look clunky. Then it looked as sleek as any sports car. The Apple II changed the world because it made desktop computing accessible. It really did all the things that everyone told you a home computer could do: write letters, balance the checkbook, draw pictures. John loved the subtle resistance of its keys beneath his fingers, the march of the cursor across the green screen.

The Apple II no more resembled the other computers on the market than a lion resembled a litter of tabbies. There was a lot of cheaper stuff out there—a choice of a Timex Sinclair 1000, a Texas Instruments TI-99/4a, a VIC 20, a Tandy TRS-80—but you got what you paid for. An Apple cost a thousand dollars or more—three months' rent. You could buy a car for that kind of money, although you couldn't travel as far in a car.

Maybe because the Apple was the only one out there built by a hacker, it was also the only one with a top that would flip open to show you the guts of the thing. It was like looking under a hood, but with an engine a hundred times cooler. You could get expansion cards for the Apple, which meant that you could build on to it, make it faster. You could double the width of the 40-column screen, making it as wide as a full page. The machine was accessible to a ten-year-old kid sitting in a quiet school office long after all the students went home. Within an hour, John had figured out how to start drawing pictures on the screen. The only sound, besides the satisfying click of the keyboard, was the movie-set loud ticking of the oversize school clock on the wall.

Sheila taught BASIC to John. He wrote simple lines of code at first. It was a solitary pursuit, but the class clown didn't tire of it. The machine was like a permanent audience. Tell it what you wanted, find the right words, and it spit back a response that was better than applause.

John lashed together his first program. He told the computer to type his name: J O H N L E E.

There it was, across the screen. Hit the execute key and see it again: J O H N L E E. Yeah. Again.

It needed something. A little flourish.

He added a bell character to the program.

So now it went J O H N L E E . . . DING!

He loved the ding. It was a celebration, of his name and of his power.

Again.

Sometimes Sheila would invite him to her brownstone in Park Slope, a part of Brooklyn that John's mom couldn't afford to live in. Sheila always let him play on the computer, and he fooled around with a program called Eliza, an application that made the computer act like a person. It was really fun to watch the computer carry on a conversation, playing the role of a Freudian therapist.

John typed: Hello, my name is John Lee.

The computer typed: HOW ARE YOU JOHN?

Fine.

WHAT WOULD YOU LIKE TO TALK ABOUT TODAY?

What's new?

WHATEVER YOU SAY.

What do you say?

MAYBE YOU SHOULD CHANGE YOUR ATTITUDE.

How?

WHAT IS IT THAT YOU REALLY WANT TO ASK?

How should I change my attitude?

IT APPEARS THAT WE ARE ALIKE IN MANY WAYS.

How am I like you?

OH, YOU ARE LIKE ME IN SOME WAY, NO MATTER WHAT YOU THINK.

Am I like a computer?

ARE YOU FRIGHTENED BY MACHINES?

Not at all.

TRY TO GET IN TOUCH WITH YOUR TRUE FEELINGS, JOHN.

How did it work? John wanted to know. Sheila told him it was one of the first programs to demonstrate something called artificial intelligence. The program could "parse" sentences, tearing the words apart and looking for key words.

HOW ARE YOU, JOHN?

Fine, I guess.

FINE? HOW SO? PLEASE ELUCIDATE YOUR
THOUGHTS.

Oh, the thoughts John could elucidate. Where would he start?
With school, where it was boring and where they were always
bossing him around? Or with the places where he'd been living?
Their apartments were always located in some of the worst
neighborhoods in Brooklyn, so dangerous his mom wouldn't even let
him play out on the street.

Or maybe the Freudian Eliza would prefer him to elucidate about
his father.

He could tell the computer about how his dad didn't live with
John and Larraine, never had. He could tell about how his dad once
brought him a hand-held pinball machine and another time took him
to an amusement park for the day. And about how everyone said John
was just like him, had the same wonderful hearty laugh.

And what would the computer say?

John could tell the computer that he hadn't seen his dad in a
couple of years, not since the time Larraine told him to stay away.
Having him breeze in with jokes and presents left Larraine looking
like the heavy. John might tell the computer that it was okay. He told
that to everyone else, told them he hardly even thought about his dad.

John and his mom moved around Brooklyn a lot. They lived in
Red Hook, by the projects, for a while. Clinton, Brownsville, Fort
Greene. In Crown Heights, they lived at the corner of Franklin and
Utica, and they wondered why so many people came in the front
hallway to push money through the slot in the landlord's door, the
same slot where they paid rent every month. Crack, somebody told
John one day. The word was unfamiliar, it was a new word in the
mid-1980s. But the tone was old, and soon they moved again.

For his birthday in sixth grade, John's mom bought him a
Commodore 64. What else? At $299, it was still too expensive.
John's mom had saved for months and he knew why—they had
talked about how to save the money, even came up with a plan. But
he still was surprised that she actually bought it. And on the morning

she gave it to him, he hugged her and yelled, "Oh Mom, I love you!"

John took a test at the end of sixth grade, and miraculously it showed that he was working at an eighth-grade level. This surprised the teachers, since John was the kind of student whose average dropped from 95 at the beginning of the year to 55 by the end. He took another test at the end of eighth grade, which he was barely passing. And guess what? He scored about as high as anybody in the whole city. It enabled him to enroll in one of the most prestigious public high schools in New York City, Stuyvesant, across the river in Manhattan. It's nothing for Stuyvesant High School to turn out kids headed to Harvard, to Yale, then on to federal judgeships. When alumnus Antonin Scalia got nominated to the Supreme Court just before John got to Stuyvesant, school officials issued a proud but understated press release. They had expected no less from a Stuyvesant boy.

What did they expect of John at school? He didn't know. He didn't care. He was there, but he wasn't. He understood everything he heard in class, but it just didn't have much to do with his real life. Drafting class was fun, because he liked to draw, and because he and the teacher had a one-joke relationship based on the teacher's willingness to do a (bad) imitation of the rappers Run DMC. But for the most part, high school just seemed like a silly diversion from John's real life on the street.

The Decepticons hung out in front of the stoic brick projects sandwiched in between crack-cap-strewn Fort Greene Park and the deteriorating Brooklyn Navy Yard. John knew that you had to belong to something, and the Decepts looked out for you so you didn't get robbed as often. The gang took its name from Saturday cartoons, a show called "The Transformers." Transformers were trucks that turned into fighting robots. John was a transformer himself. He transformed into a Stuyvesant student every morning. He transformed into a street kid every night.

John didn't think it was a really tough gang, because they were not into drugs. And hardly any murders had been pinned on them. Decepticons were just into the usual stuff, really. You hung out in front of the bodegas in the neighborhood. You told girls, "Hey, I'm a Decept." Or if you met someone, and he said, "Who are you, man?" you said, "I'm a Decept," and he said, "I used to be a Decept, too." Then you knew he was OK. That's how it worked on the street.

It was weird trying to integrate the two components of John's life—street life and school life. They were like two halves of a life, and they allowed him to have different personalities. One was gregarious, quick-witted, and funny, helping some kid at school learn to use a computer. Another personality got a rush from thinking about how a counterman looked when you walked into his bodega and told him to empty the cash register.

John didn't meet a lot of Decepts at Stuyvesant. He didn't meet a lot of calculus students under the broken streetlights at the Raymond Ingersoll Houses. John never once doubted that he'd go to college. He thought, I can pull a robbery, go to jail for a year, and then go to college. Sometimes he didn't know who he was.

One lesson that stuck with him: John and a friend were on the J train in Manhattan, near the Delancey stop under the Lower East Side. The tracks are dug under the Bowery, where the bums used to loll around the pawnshops and cheap-beer bars. John's friend got into a staring contest with some guy across the aisle in the train, one of those situations that everyone except teenage boys tried to avoid.

"What are you staring at?" the guy across the aisle blurted.

"Absolutely nothing," said John's friend.

Then the guy across the aisle shoved his hand into his waistband, as if he was about to pull out a gun. John's friend jumped up and walked into the next car, then all the way to the front of the train until he found a transit cop. The train was clattering and rocking, the lights blinking on and off as if there was an earthquake. He brought back the cop, pointed, and said, "This guy says he has a gun."

The cop frisked the indignant rider, who was clean. The cop glared at John's friend, then left the car. At the next stop, John's friend followed the unarmed guy into the station and started beating on him.

John heard his friend say, "What you say you got a gun for, man? You would've killed me if you really had one. Next time you better have one."

Here was a lesson more relevant than anything John learned at Stuyvesant. Authority figures were just units in the great system that is life. They were individual lines of code, and if you used them in the right way, you could get a predictable response. Power was something you could borrow.

John knew a place where it was easy to get money. Back by this

Brooklyn street mall, Albee Square, a raucous place where hawkers press half-off come-ons into your palms. Fayva Shoes. Burger King. A Stetson hat store that now mostly sells Kangol caps. Cheap gold ID bracelets, your initials in crass block letters. Behind long collapsible tables, solemn Muslims robed in white sit burning incense, selling fragrances in colored vials.

Nearby, under the ramp to the Manhattan Bridge, was a whorehouse, a full-service whorehouse where working men could cash their paychecks and, if they were lucky, even come home with the change.

If they were unlucky, they ran into a Decept with a gun. John had one, a small .25-caliber pistol with a tiny clip. He got it from a friend who brought it from California, along with a sawed-off pistol-grip shotgun. John and a partner would rob the men coming out of the whorehouse. They would pick a guy wearing a wedding ring. What was he going to do, call the cops? "Officer, I was stuck up while leaving this *whorehouse*." No way. John and his partner would approach quietly, they didn't want to scare him, and then John would jump right in his face and say, "Give me all your money right now!" Sometimes the guy was too scared to hand over his money, and John would have to pat him down, take his wallet.

One cold winter night, they stuck up some married guy and what did they get? Pesos. They were looking at the pesos, wondering what the hell, when a police car pulled up, and they were busted.

John's mom was very upset, even after he told her that he found the pesos on the street and was just asking the guy how much they were worth. The judge wasn't buying any, even though the victim didn't show up to press charges. The judge looked at John standing before the bench, and John could tell he didn't believe the pesos story.

Here's what the judge said: "A Stuyvesant High School student has no business being involved in that situation."

For some reason, John started to think that day about the computer that had been sitting in his bedroom since sixth grade. A Stuyvesant student shouldn't be in a situation like this. It wasn't leading anywhere, not anywhere he wanted to go. He knew now that he couldn't control things on the street. He was just a component there.

John got an idea. He knew this girl who had a crush on him. Having only twenty dollars, John hit her up for sixty dollars more, and sent her into an electronics store on 42nd Street in Manhattan. The blocks here were thick with these stores, which all had strobe-lit windows displaying last year's whole Sony Walkman line, miniature TVs, row after row of cameras, palm-size video games, answering machines, everything a guy with twenty dollars might lust for. These were the kinds of stores where you had to know what you wanted before you went in, and know your price. John knew both, only he was acting kind of goofy. "You go," he told the girl, because secretly he was afraid the salesclerk would ask a question he couldn't answer. He didn't want to look ignorant.

He waited outside, pacing a little in front of the window.

The eighty dollars was enough to buy a modem for John's Commie 64. He didn't notice later when the girl got bored and went home. He was too busy trying to connect it.

Why was he so intent? He wanted to be a hacker. He didn't even know why. All he knew was that it was way more interesting than hanging out on the street, robbing people.

On the first day he managed to connect to a bulletin board, a chunky-crude picture of a bear—the growling face of a bear—filled the screen in John's bedroom. He had called Papa Bear's Den, and with a goofy name like that, you know no one from the Legion of Doom was hanging out there. It was more the kind of place where a data processing joe from the post office might log in to feed his computer hobby.

But to John, it was cool. He had actually figured out how to get his modem to call this number, figured out how to connect to another sentient being, just by using his computer.

He learned to post electronic messages:

```
I'M INTERESTED IN BECOMING A HACKER.
SIGNED,
DAMAGE
```

Now he had his first hacker handle. Damage. It was the first word he thought of when he was typing. It was an honest-to-God nickname, a passport in cyberspace.

This was so cool, this bulletin board stuff, that he had to share the experience. He found a friend in his neighborhood, someone who, of course, didn't have a computer. He invited the kid back to his house. John activated the modem, and they heard the clicks as it dialed Papa Bear's Den. The image of the growling bear face filled the screen. John's friend just looked at the bear, then at John. He thought John was ridiculous.

Of course, the friend had a very different reaction the next time John invited him over to play with the computer. That day, a few months later, John logged into TRW's national database of credit records and started to read people's credit histories. No argument, that was cool.

By that time, John had abandoned his old handle. He was calling himself Corrupt now. That had been his street name. It should do just fine in cyberspace.

John educated himself so fast that within a few months after he bought his modem, he was on track with the other MOD boys. For one thing, John figured out that some rules are the same, whether you're on the street or in cyberspace. If you want to get ahead, no one is going to just *let* you. You have to take what you want and get there yourself. He played a little game sometimes. He called it Let a Hacker Do the Work. Like the time he called a hacker named Signal Interrupt in Florida, and sweet-talked the kid out of all *kinds* of information, just by claiming to be a member of the Legion of Doom.

Another way cyberspace was like the street was that it helps to have friends.

John hung around on a German chat board with some hackers who called themselves the Eight-Legged Groove Machine. One of the four members had been busted in England, so there was an opening for two legs. Now they were teaching John how to crack VAX machines.

The hacking scene is really big in Europe, where, in fact, it's not even against the law for kids to log into private computers and look around. That's why so many European hackers cross the Atlantic to dig around in U.S. computers. Where they come from it's like a sport. Lots of European bulletin boards and chat systems are popular with U.S. kids because the quality of information traded online is so high.

How did a guy from Brooklyn get on a chat board in Hamburg, Germany? Some philes on local bulletin boards had told John about Tymnet, a private network that businesses use to connect their

computers to one another around the world. It's like a phone system
that computers, not people, use. Every major company, it seems, is a
Tymnet customer and communicates on Tymnet's pervasive network
of computers and phone lines. On Tymnet, Johnson & Johnson
operates a sub-network to link its computers in different cities and
countries. So does the Bank of America. And so does TRW. Even the
federal government is a Tymnet customer and uses the network to
link sensitive National Security Administration computers to one
another.

If you're a Tymnet customer, it's easy to use the network to call
another computer anywhere in the world. You just dial a local phone
number that hooks you to the system, then type in your user name
and password. Theoretically, the password keeps the system secure
from unauthorized intruders.

But John learned that there were all sorts of ways to circumvent
the security to get inside the Tymnet system. All you had to do was
call Tymnet's local phone number, and then when the modems
connected, simply type a certain NUI, which stands for Network User
Identification. Then you type in a password. The philes told John
what NUIs and passwords to use.

But John still hung out on American bulletin boards, too. On
Phuc the Pheds, he met this guy from the Bronx who has the same
sense of humor as John. He's younger, but he seemed pretty cool. His
name was Julio Fernandez, and he called himself Outlaw.

Julio really looked up to John. That's because in the years since
his sixth-grade teacher gave him a book that introduced him to
computer programming and games, Julio hasn't met too many
people who knew more than he did about computers. John definitely
knew more. Julio spent hours every day trying to break into other
people's computer systems—and so did John. John always had
suggestions of new things to try, a command that might work, a back
door that nobody else had thought of. Julio, who was only fifteen,
lived in a fantasy world, where the computer underground was an
exciting place, and outwitting the law was something a cool hacker
could do forever. Julio's mother encouraged him to stay in the
apartment and play on his computer. She figured that it was
dangerous to play outside on the streets of the Bronx; her son was
safer at home. What could be dangerous about sitting at a keyboard
and typing all day and all night?

* * *

John would log in to Sage pretty often, so it's no wonder Eli noticed him there. For a while, after that first phone call, John developed a telephone relationship with both Eli and Mark. They traded mutually beneficial information. They get along because they see things the same way.

John starts to think about joining up with MOD later in the fall. Somehow, he'd been bounced out of Stuyvesant, because things didn't work out, and he ends up at City-As-School.

One day, while he's hanging around out in front of school, break-dancing, dreadlocks flying, John leaps up into the air, straddling, and jumps right over another kid's head. Like he's flying.

He lands and turns, as a skinny white kid watches his moves. The kid has dark hair, meticulously groomed. The kid is Mark.

And now, the two know each other outside the confines of cyberspace. They meet from time to time at school, on the days when they bring in their journals to show their advisers. John carries around a lot of heavy books on VAX/VMS programming, and Mark likes to see that. It shows that John is serious.

SEVEN

Paul went to college after the summer ended, to Polytechnic University on Long Island. It was only about an hour east of Queens, and not such an odd choice for the valedictorian, because the school offered him a Presidential Scholarship. The only other school he'd considered was Queens College. He didn't want to go too far away from home.

But Paul often wondered if he's made the right choice. He thinks it's a depressing campus, with so many engineering students plunked down on this isolated island of square, modern buildings. The school isn't in a college town, no bookstores or pizza hangouts or coffeehouses to walk to. No revival movie house for whiling away the weekends. The campus is on the edge of a town called Farmingdale, which is just a typical middle-class suburb, with a bank and a stationery store and a hardware shop. Not much of a lure, and it's a long walk. So once Paul was at school, he was stuck. Neither Paul nor his mom had a car. Paul got sick of playing Ping-Pong and foosball in the basement of the dorm. He went home as many weekends as possible.

One Friday during the fall, Paul rode home from school on the Long Island Railroad, looking forward to marathon hacking sessions alone in his basement. Although he was sharing information with the other MOD boys, he hasn't told anyone about some of the systems he explores. He's noticed that soon after Eli and Allen get into a computer, it closes down. Not that they crash it, necessarily. But they're careless. Paul thinks that after a week or so, system administrators get wise to the trespass and lock out all intruders. So Paul was keeping a few Unix systems just for himself.

On the train ride from Farmingdale to Queens, he sits by the window. Monopoly tract houses whiz by. Town squares that got swallowed by six-lane roads and fast food. A Japanese grocery store in a strip mall. Five-story office buildings crouching on lots barely big enough for the houses they displaced. Gas stations and cars waiting in traffic. And now, he notices the telephone poles and the thick black cable strung between them, and the cables to the buildings and to the gas stations and to the homes.

Everyone is harnessed to that electronic tether. These people are all under my control, Paul thinks. And it feels good to be so powerful.

But there's something that nags him, too. He'd told Eli and some of the other boys about one system that he should have kept secret. It's called The Learning Link. Hac, who had found The Learning Link in the first place, had warned him not to give the phone number to anyone else. But Paul eventually did.

It was hard not to.

Eli and other MOD members had been complaining that they didn't have enough systems. And Hac had pretty much dropped out of the scene and into the Marines. After he'd gone to boot camp on Parris Island, the last thing he'd be thinking about was hacking. So Paul didn't see the harm. He felt as though he should give them this Unix because it wasn't that sensitive, at least not compared to a phone company computer. Besides, they were all in this adventure together.

What the adventure was and where it was leading was unclear. The whole point, for Paul at least, seemed to be to satisfy a craving to explore new computers, new territory. The world was full of computers, glorious unknown computers crammed with information, each exhibiting its own permutation of programming language and full of its own specific mysteries.

All you had to do was find them.

Paul found lots of computers back when he was scanning 800 numbers. Allen had stumbled on the Eye Center computer by accident. For months now, the MOD boys had been looking for uncharted systems. That's what hackers do; they're always scouting for new properties. In a way, they're just like horse traders. If you're a hacker, you swap computer systems. What's a really good Unix worth? Two VAXes? A switch dialup? Is it AT&T or a local phone company? To be a hacker in the late 1980s was to be a kid with a notebook stuffed with passwords for Unixes and VAXes, switch dialups, and all kinds of university mainframes. Of course, the passwords don't last forever. So the hackers needed new horseflesh. So Paul gave them The Learning Link.

But Eli and the others aren't under my control, Paul thinks.

The Learning Link is one of the first great experiments in networked education. Owned by Channel 13/WNET, New York City's public broadcasting television station, the network is like a collection of bulletin boards. Originally, it was a way for teachers to order rebroadcasts of educational shows. Say a science teacher wanted to show a Nova episode. She merely requested it through The Learning Link, and if WNET could handle it, the station would transmit the show during dead air time, maybe at three A.M. That way, the teacher could record it on a VCR.

The Learning Link has evolved, like everything in the networked world, into a novel way to communicate. Educators and librarians can send electronic messages to each other over the system. The Learning Link is extremely useful for thousands of schoolteachers and students throughout New York, New Jersey, and Connecticut. Educators post announcements and participate in discussion forums. For the teaching community, it's a vital link. A school has only to buy a subscription to the system.

Non-subscribers can log in and look around the system with limited privileges, simply by signing on as GUEST.

For more than a year, ever since Hac saw The Learning Link's phone number advertised in a brochure at Flushing High School, Hac and Paul had been doing just that. They were quiet about it. No one even knew they were there.

For Eli and other MOD members, though, The Learning Link is a new conquest, and they concentrate on wresting control of the system. On the surface, it looks like one of those computers that seems largely uninteresting, just a Unix computer, no more, no less. It doesn't run any particularly exotic flavor of the Unix operating system; there isn't anything intrinsically interesting in the data stored in the system. From a veteran hacker's standpoint, there is nothing to be learned here. If Eli had a lot of other systems to explore, he might not even log in to The Learning Link.

But Eli is enchanted by The Link's connection to the mass media. If the media notice you, you become famous.

Earlier this year, the media had latched on to a story about the trial of a hacker named Robert Morris, a Cornell University student who unleashed a rogue computer program that shut down the global web of computer networks known as the Internet. It was the first time the hidden world of cyberspace, the hidden world of computer hackers, surfaced. People realized there was this new, uncharted universe of activity out there, and they wanted to know more. Oh, the things the MOD boys could tell them. If only someone would ask.

Eli couldn't stay away from the WNET computer.

Now, in November, MOD members have gotten root access to the system, meaning that they can exercise complete control over The Learning Link. Root access means you are a super user. You have the power to execute any command you type. You have the power to look at any file in the system. You have the power to modify the system in any way you choose. It's not so hard to figure out how to get root access on a system with security so lax that any wanderer can log in as GUEST. These were the days before people started locking their doors in cyberspace. Besides, why would anyone want to crack a totally benign educational network that attracts public schoolteachers and guidance counselors?

There are any number of ways to crack an unprotected Unix computer. You might find a hole that allows you to write a program that you could hide in the system. When such a program (aptly known as a Trojan Horse) is unwittingly executed by the system administrator, it adds a new password file to the system—enabling you to log in as an authorized super user. Once you're a super user, you can create and delete user accounts. You can read anybody's mail. You can shut the whole thing down.

Why would the MOD boys want to shut The Learning Link down? The more systems they "own," the more power they have, right? But here was the catch. Nobody would know they have the power unless they advertised.

To this day, Eli swears he never meant to crash The Learning Link that November evening. He said he and another MOD member were just exploring. He certainly had never crashed anything intentionally. (Plik.) Of course, once when he was fooling around on a private computer, he accidentally deleted a critical command that the system administrator needed to keep track of what everyone was doing on the system at any given time. That was almost as bad as crashing a system. But that time, he got Paul to fix it.

Since it's almost Thanksgiving, Eli just wants to leave a greeting card. He wants to let everyone at PBS know he is there. Let the media know that the most elite dewds in cyberspace live in the five boroughs, right under their noses, if anyone would care to check. The way Eli looked at the world, all the media wanted is a sensational story. MOD was a sensational story, and nobody even knew it.

Eli bypasses Hallmark and writes his own greeting, electronic style:

```
Happy Thanksgiving you turkeys, from all
of us at MOD:
The Wing / Acid Phreak / Scorpion /
Supernigger / Nynex Phreak / HAC / Phiber
Optik / Thomas Covenant
```

Eli sends a command to The Learning Link computer, ordering it to type out that message on its printer. The printer will print the message over and over again, until paper spills out all over the floor. Now everyone at Channel 13 will know that they've been hacked by the best. It's just a spoof, a goof, a gag. Eli means no harm. He's just a kid. In his mind, it's like tying someone's shoelaces together to watch them trip.

But something else altogether happens in the early morning hours of November 28.

Eli would later tell a judge that some other MOD member took the message that Eli sent to the printer and modified it. Then, Eli would say, that person rigged the system so that whenever anyone logged in, instead of seeing the familiar Learning Link greeting, the user got Eli's greeting card. For good measure, the system administrator's password was changed, so that the real system administrator couldn't run the system or erase the message. Like being locked out of your own house.

In fact, no one can get past the message. Every schoolteacher in a three-state area is locked out. And reading the graffiti that the MOD boys composed is not making them any happier. The message says:

```
Haha! You want to log in? Why? It's
empty! HAHAHAHA!
MOD brings Seasons Greetings to llwnet
(Channel 13)...Happy Thanksgiving you
turkeys, from all of us at MOD:
The Wing / Acid Phreak / Scorpion /
Supernigger / Nynex Phreak / HAC / Phiber
Optik / Thomas Covenant
```

Employees of WNET must rescue the system by reconstructing it from backup computer tapes they keep for emergencies.

Eli would later tell a judge that he was not responsible. But the very next day, the DNR on Eli's line notes that Eli calls The Learning Link again. And soon after, the whole system crashed again. Someone had erased all the files.

Eli said it wasn't him. He said when he found out The Learning Link system had crashed, he got a bad feeling in his gut, as if he were walking down the street with a friend and the friend said, "I'm going to trip this lady coming toward us." And then he trips the lady!

Was this what real hackers do?

A day or so later, Allen and Eli have a three-way phone call with Paul at college. They tell him what happened.

Paul feels kind of sick. For some reason Paul can't figure out, Allen and Eli are laughing about it. Even Paul laughs a little bit at first, although he can't explain why. But it makes him uneasy.

Paul knows they went too far. Heckling your peers is one thing. Even scanning calling card numbers is okay, because it seems victimless, and how else can you afford to call out-of-state computers? But deliberately crashing a system that hundreds of people (hundreds of *adults*) depend on every day is, frankly, kind of stupid.

He thinks about the hacker ethic, and the conversation he had so long ago with Hac—Hac!—about the difference between hackers and crackers. The lines were starting to blur for Paul. Hackers are only trying to learn how things work: Thou shalt not destroy any system.

But on the phone to Eli and Allen, Paul says only this: "It doesn't seem like such a good idea."

Now comes a weird coincidence.

On the day after the Learning Link crash, on the twenty-third floor of New York Telephone's midtown skyscraper, Tom Kaiser performs his morning DNR check. He looks down the list of calls made from Eli's house in the past twenty-four hours, and he comes across a phone number he's never seen before. It's a Manhattan number, and he looks it up in his directory. He sees that the phone number is registered to Channel 13, the PBS station. That's odd.

Kaiser is about to call WNET when his phone rings.

There's a woman on the other end of the line, and she's really upset. She tells Kaiser that she works at a place called The Learning Link. She didn't know who to report this to, but the phone company's business office gave her Kaiser's name in Security. She tells him what happened and asks Kaiser to trace the calls.

She wonders why anyone would do something like this.

Kaiser desperately wants to say, "My God, what a coincidence! I was just about to call you!" But he can't, of course, because he can't compromise the confidentiality of his investigation. All he can do is listen to her story, and then, at the end of it, say gently, "I know someone you can call."

He gives her the phone number of Secret Service headquarters. The Secret Service is very cooperative, someone gets all the details of the crash, and tells the woman the agency will investigate. And that's how the Learning Link crash would become the pivotal event in the case that the federal government was slowly building against the boys in MOD.

Were it not for this confluence, the Learning Link incident might never have been singled out. But as it happened, the system crash was something that law enforcement could understand immediately. It was not as ephemeral as trying to grasp the definition of a switch. It was concrete. The system had been working, the system was now crashed. An organized gang of hackers did it. It was just the event to catapult the MOD case to the top of the pile in the Secret Service's telecommunications division. Secret Service agents started having frequent conversations with prosecutors in the U.S. Attorney's office. They made plans to execute search warrants at the hackers' houses. The Learning Link crash was, in fact, the event that four years later would cause a prosecutor to smile wryly, as he explained, "This made the case sexy."

Even though Eli's missive to the media never got delivered, the general message was getting through. Hackers are out there, people, and we'd better figure out who they are and what they're doing. The Learning Link affair was remaining a secret for now, but thanks to the Morris story, newspapers and newsmagazines are assigning reporters to learn the new definition of the word *hacker*.

In that spirit, *Harper's* magazine decided to host an electronic forum on the topic of "Is Computer Hacking a Crime?"

The forum took place in cyberspace itself, during an eleven-day period in December. The gathering place, the little corner of the electronic world where the forum rambled like a never-ending cocktail party, is called the Whole Earth 'Lectronic Link, better known as the WELL. The San Francisco Bay Area–based commerical bulletin board system is a seminal outpost that attracts some of the best minds of the communications revolution. On the WELL, an explosive mixture of computers and communications was being brewed.

About forty guests log in from their homes and participate in an open-ended, freewheeling debate. They were an eclectic group, and among their number were Eli and Mark (uninvited) and a retired cattle rancher and lyricist for the Grateful Dead, John Perry Barlow. Given the amount of time they spent there, it was bound to happen that the boys and Barlow would cross paths online one day. But who could have imagined the repercussions?

John Perry Barlow tells everyone he lives in Pinedale, Wyoming, but that's not really true. He actually lives in a much more isolated place, a place called Cora, pop. 4. Cora makes Pinedale look like Los Angeles.

Barlow's grandfather founded Sublett County and built a cattle ranch next to 15,000 acres of forest service land. Barlow's great-granduncle was the first white man to winter on the upper headwaters of the Wind River.

Barlow went east (well, east to Colorado) to prep school, where he was schoolmates with future Grateful Dead band member Bob Weir. After Wesleyan College and a stint in New York City working on a never-published novel, Barlow wrote songs for the Dead before meandering home to the Bar-Cross Ranch. That's − + if you're branding calves, and as Barlow says, "They sure got the minus sign in the right place."

Barlow traveled many trails in his youth and by the mid-1980s, with the ranch's fortunes in decline, he needed another way to make a living. He turned to the thing he knew best, writing. He bought a computer, a Compaq thirty-pound luggable, mainly because he figured it would be easier to use than a typewriter, "a better Wite-Out." But in 1987, Barlow learned his computer was also good for something else.

That was the year he discovered that the futuristic place described by science fiction writer William Gibson as cyberspace was in fact here. A friend, a fellow Deadhead, convinced Barlow to buy a modem and log in to the WELL to communicate with other Dead fans. Right away, Barlow saw the similarities between this new electronic community and the small towns of his youth. Small towns were disappearing, but new ones, a generation of computer-connected villages, were replacing them.

And the beauty of it was, just like in small town life, everyone knew everyone else there. People worked together, shared information, and pooled their knowledge, creating a vast collective consciousness. It was a primitive settlement, people were still learning how to communicate. There were no rules or laws yet. Instead of physical possessions, all people carried was knowledge, which they shared. Barlow moved right in and started writing about it.

By now, two years later, Barlow was spending so many hours a week on the WELL that he was a beloved "net personality." So it was no surprise that he was asked to participate in the Harper's Forum.

There Barlow met the two gate crashers, hackers who showed up uninvited and who called themselves Acid Phreak and Phiber Optik.

When they logged in to the forum early in the evening of Day One, the conversation's tenor shifted radically. Gone was the high-minded, theoretical discussion about the rights of privacy versus the right to explore. In its place appeared a challenge born of fluctuating hormones and adolescent invincibility:

ACID PHREAK: There is no one hacker ethic. The hacker of old sought to find what the computer itself could do. There was nothing illegal about that. Today, hackers and phreaks are drawn to specific, often corporate, systems. It's no wonder everyone on the other side is getting mad. We're always one step ahead.

Even as he typed, Eli was defining himself, creating his own new hacker ethic. It was a philosophy in which exploration for the sake of discovery is its own justification. And it was also a philosophy in which he saw himself as the baddest gunslinger to ever ride into town.

But no hint of that teenage transformation got conveyed to the rest of the law-fearing, job-holding adult participants. They didn't even know who this Acid Phreak character was. They certainly had no way of knowing he was just a teenager with a poster of a bikini-clad model on the closet door in his bedroom. They didn't know that it had been a blind kid named Carlos, who lived across the street, who'd taught Eli the ABCs of phone phreaking. (Carlos's mom was from Argentina and Eli's mom was from Costa Rica and the two were friends. So it was only natural that Eli and Carlos would be friends, too. Carlos even had a computer, which, of course, he couldn't see. Later, Eli got his own computer, and learned enough to turn the pay phone down the block on Parsons Boulevard into a free phone. It was a kick to walk by the phone and see a line of people waiting to call home to Colombia. And he'd think of Carlos, and how much a guy could do if he put his mind to it.)

Of course, all the forum participants really needed to know was that the *Harper's* forum on hackers had flushed the elusive species from the bushes.

Barlow himself could well remember his own trespasses, including a late-night climb over a government defense installation's fence. In finding himself infatuated by the impulses of Acid Phreak and Phiber Optik, he was not alone.

Within hours, the forum's participants had declared their fascination with Acid Phreak and Phiber Optik, and by the afternoon of Day Six, the two hackers had become celebrities. Even their most outrageous adolescent crowing and foomphetting was given careful consideration.

This emboldened Eli, who espoused the view that any computer system vulnerable to hackers was fair game. It was the system administrator's fault if the computer got hacked. It was the administrator's responsibility to keep it secure, after all. If he was too dumb to do that, then hackers had the right to trespass. It was the old Eye Center argument again.

Did Eli really believe that? Once this philosophy was posted on the WELL, he couldn't take it back. It became his position. In a matter of weeks, it was going to be printed in *Harper's* and then every subscriber in the country would read it.

One participant questioned Eli's assertions, posting that computer networks "are built on trust. If they aren't, they should be."

ACID PHREAK: Yeah. Sure. And we should use the "honor system" as a first line of security against hack attempts.

This cocksure disregard for honor and honesty struck a chord with the very hacker-gawking adults that Eli had charmed earlier. As Barlow later wrote, "Presented with such a terrifying amalgam of raw youth and apparent power, we fluttered like a flock of indignant Babbitts around the Status Quo, defending it heartily." One participant, a former hacker named Jef Poskanzer, shot back this sarcastic reply:

JEF POSKANZER: This guy down the street from me sometimes leaves his back door unlocked. If I had the chance to do it over, I would go in the back door, shoot him, and take all his money and consumer electronics. It's the only way to get through to him.

ACID PHREAK: Jef Poskanker (Pus? Canker? yechh) Anyway, now when did you first start having these delusions where computer hacking was even *remotely* similar to murder?

Before publication, *Harper's* edited the conversation, deleting Eli's name-calling and thereby denying readers a glimpse of the adolescent hiding behind the fearsome nickname. In fact, editing eliminated any flagrantly immature statement that Eli made, rendering him a larger-than-life electronic desperado. It's a one-dimensional portrait, but one that he will find hard to resist in the coming months.

The media and the monster. Which is the creation and which the creator?

The forum participants bristled at the bad-boy attitude Eli sported.

BARLOW: Acid, my house is at 372 North Franklin Street in Pinedale, Wyoming. Heading north on Franklin, go about two blocks off the main drag before you run in to a hay meadow on the left. I'm the last house before the field. The computer is always on. . . . You disappoint me, pal. For all your James Dean-on-Silicon rhetoric, you're not a cyberpunk. You're just a punk.

ACID PHREAK: Mr. Barlow: Thank you for posting all I need to get your credit information and a whole lot more! Now, who is to blame? ME for getting it or YOU for being such an idiot?!

The language was heated now, and in cyberspace, the words themselves are all you have to guide you. The words aren't tempered by inflection, or by a wry smile that says you're really just kidding, or by body language that says you're more amused than enraged. No, the words are the whole communication, and they sure could look stark on your computer screen. On Day Ten, Barlow was writing what can only be called fighting words.

BARLOW: Let me define my terms. Using hacker in a mid-spectrum sense (with cracker on one end and Leonardo DaVinci on the other), I think it does take a kind of genius to be a truly productive hacker. . . . With crackers like Acid and Optik, the issue is less intelligence than alienation. Trade their modems for skateboards and only a slight conceptual shift would occur.

Until now, Mark had read more than he'd written, leaving the boasting to Eli. But now, he was incensed. He couldn't take it anymore. And so the gunslinger, the Fastest Two Fingers in the East, typed this fateful response at 10:11 P.M. on the evening of the Tenth Day:

PHIBER OPTIK: You have some pair of balls. . . . Hmm . . . This was indeed boring, but nonetheless:

And then he did it. He rendered this theoretical discussion absolutely real. Mark posted a copy of Barlow's private credit report, culled from TRW, right smack in the middle of the forum for all the participants to read.

Barlow, sitting in his office in Cora (post office Pinedale), watched the intimate details of his finances appear before him on the screen.

No matter that the snippet from TRW was abbreviated, and incorrect. No matter that reading credit histories, and then copying the files into his own computer, was far different from actually having the ability to alter a report. No, none of that mattered. For Mark had just demonstrated that some hackers have very real power over the lives of everyday people.

Barlow wondered what life in America would be like without credit.

As Barlow later wrote: "I've been in redneck bars wearing shoulder-length curls, police custody while on acid, and Harlem after midnight, but no one has ever put the spook in me quite as Phiber Optik did at that moment."

Barlow asked Phiber Optik to call him.

Of course, Mark did. And there, well beyond the spotlight of the electronic forum, the two met, man-to-man, on the phone. The romantic and the outlaw.

Barlow came to see Mark as a "pencil-necked kid with the same desire to violate the forbidden that has motivated male adolescents since the dawn of testosterone."

The phone conversation burgeoned into a telephonic friendship, and Barlow came to believe Mark is "surprisingly principled," with "impulses [that] seemed purely exploratory."

Mark is Mark, and in explaining his philosophy, that of a lone scientist in his laboratory, he never saw the need to fill Barlow in on the less savory aspects of MOD's adventures. Whatever Eli or other MOD members did to The Learning Link, they did on their own, without Mark's help or commiseration or even knowledge.

Mark was articulate. Mark was earnest. And Mark was nothing if not persuasive. Mark would make a good poster boy for any group that decided to carve out specific civil liberties for hackers in cyberspace.

Barlow wasn't thinking about all these things, not yet at least, when he had his first phone conversations with Mark. The rest would come later.

For now, Barlow began to wonder "if we wouldn't also regard spelunkers as desperate criminals if AT&T owned all the caves."

EIGHT

January 15, 1990, seems like just another routine day.

On the day that AT&T's system will catastrophically fail, Tom Kaiser commutes on the Long Island Railroad into Manhattan. While he tries to read the paper, he feels the usual early morning anxiety that comes from not knowing what the hackers were doing the night before.

When he arrives at his office at about 8:30, he immediately checks on the black box. The DNRs show the hackers had a busy night, but nothing appears out of the ordinary. Cat and mouse. But isn't the cat supposed to win?

He updates Staples, who moved back to his own office after the strike ended, on the previous evening's hacking activities. On the phone they go over it together, and nothing they notice, nothing they talk about, sends an alarm through their heads. Nothing on the DNRs causes either of them to say, "Oh, my God," pick up the phone and call the authorities.

By midafternoon, however, it is clear that today will be different. Kaiser hears the news first on the radio: AT&T has crashed. Journalists are speculating about the damage. Will the crash spill over to affect other crucial services? Will telephone transmissions at airports be blocked? How long will AT&T customers be cut off from communicating with relatives—aging parents, for instance—in other parts of the country? Meanwhile, hundreds of customers are calling AT&T, demanding an explanation. Hundreds more call New York Telephone, unaware that the local phone company has nothing to do with the long-distance glitches. The average customer doesn't understand and doesn't care about the early 1980s' divestiture that separated AT&T from the local phone companies. To the average customer, the only truth is this: if a phone call isn't going through, then The Phone Company is to blame. Simple. Now fix it.

At his battered metal desk Kaiser listens to the news and all he can think is, Did the hackers do this? And worse, he thinks, did *we* let this happen?

Almost every day for months, Kaiser and Staples have played the "what if" game. What if a hacker makes a mistake and crashes the computer? What if a hacker deliberately sabotages the computer? What if. Every workday has been a game of Russian roulette for the New York Telephone investigators. The hacker's gun is loaded with bullets, and every day the investigators have wondered if he'd shoot one into the brain of the phone company.

And now, AT&T is down. It could be mere happenstance that the nation's first long-distance phone crash was occurring right in the middle of Kaiser and Staples' biggest computer intrusion investigation ever. It could be one of those coincidences that dog Kaiser. But he doesn't like the odds.

Within moments, he and Staples are on the phone again, hashing out possibilities, figuring out what to do next.

"The hackers haven't done anything obvious," Kaiser says, looking at the DNR menu on his computer screen. He scrolls slowly through each of the phone numbers that the hackers dialed in the past twenty-four hours.

"The initial news reports say AT&T has code problems," says Staples. They both know that doesn't let the hackers off the hook. "Code problems" is a simple way of saying there's some internal snag in the AT&T operating system, some bug that may have been introduced into the millions of lines of computer code that run the mammoth network. It's the equivalent of saying that somewhere in the world, in one copy of the Bible, it says "Jezus." How do you find the word, in one line in one page in one copy of the book? For AT&T's technicians, the awesome task ahead would be to isolate the affected area of the code. Find the bug. Maybe there's a misplaced command. Maybe somebody forgot to hit the RETURN key when typing a line. Maybe it was just a few incorrect keystrokes that upset the delicate balance of the entire operating system. Of course, if the whole phone system crash was caused by a bug, that raises another question. How did the bug get into the code in the first place?

If it was deliberate, then it was malicious. That could point to hackers. If it was a mistake, then it *still* could point to hackers, because if a bunch of kids were fooling around in the system, who knows what they might type in ignorance? Even if no obvious connection pops onto Kaiser's screen as he scrolls through the DNR list, he has to wonder. Maybe the crash was precipitated by a hacker or hackers that Kaiser doesn't even know about. He's long suspected that there might be others in the phone company computers, wilier intruders who covered their tracks so well that Kaiser never suspected their presence. And what about all the phone numbers that Kaiser's hackers call from their home phones? Could one of those numbers connect the hackers to a ringleader, some mysterious and dangerous instigator whom Kaiser doesn't even know about? You can "what if" yourself to death in some jobs—and Kaiser has one of those jobs.

"Did we make a mistake?" Kaiser asks Staples. "Did we make the right call in which hackers to follow, or is there someone out there who's more important, and we aren't following them?"

But they could indulge in only so much self-flagellation. It hadn't been their idea to drag out this case for so many long months. After all, they've already turned over enough information from the DNRs to enable the prosecutors to apply for search warrants at the hackers' houses. They even passed along the tip about The Learning Link to the Secret Service. It was up to the government now to get the search warrants. The Secret Service and the U.S. Attorney's office kept telling Kaiser that it was in the works. Any day now, they'd say.

At Secret Service headquarters downtown, a certain special agent named Rick Harris was torturing himself with the same questions that were hounding Kaiser and Staples. Harris ran the new tele-communications division in the agency's Manhattan field office. By virtue of that title, he'd been monitoring this case for months, getting regular updates from the agent in charge of the case. Despite the impatience of Kaiser and Staples, the Secret Service had been working steadily on this case. It was an opportunity for the agency to set a precedent in a new area of investigation. You see, the Secret Service had been slowly investigating more and more computer crime cases. Ever since a new federal law in the mid-1980s had given the Secret Service authority to investigate credit card fraud, among other crimes, the agency had been gearing up for the electronic era in crime. Credit card fraud was a natural digression for the Secret Service. The agency is the law enforcement arm of the Treasury Department, after all. More and more credit card fraud was being perpetrated in cyberspace. You didn't have to be Bill Gates to know how to use your computer to log in to an underground bulletin board and find lists of purloined long-distance calling card numbers you can "use." Of course, in that instance, you'd also be using the telephone lines to commit a crime.

Computers. Telephone lines. Whose jurisdiction was this? It gets complicated out there on the cutting edge of electronic crime, and that was why the Secret Service established a telecommunications division in its Manhattan office in 1989. The division, under the leadership of Harris, was supposed to figure out exactly what electronic crimes were occurring and how, and—most important—how to fight back. This was a high-profile assignment for an agent still in his thirties. Harris, who has likable, laughing eyes and the easy shamble of a sheepdog shaking off sleep, believed from the beginning that this area of law enforcement was on the verge of exploding.

His first cases showed him just how pervasive electronic crime had become. The agency was in the middle of investigating several cases involving call-sell operations. A "call-sell" is, simply, a situation where someone illegally sells use of a phone to someone else. You see it all the time in New York City—long lines at a street corner, with recent immigrants waiting their turns to call home to the Dominican Republic or Colombia for five bucks.

Here's how a call-sell operation is set up: First, the mastermind breaks into a PBX. That's a Private-Branch Exchange computer that manages phone calls internally for big corporations. A PBX is just like a mini-switch, handling all the phone traffic in and out of the corporation's offices. By hacking a PBX, an intruder can get control of the system, and then dial out on one of the organization's phone lines to anywhere in the world. In the freewheeling experience of New York

City, where 85 percent of all PBX fraud occurs, street entrepreneurs advertise the availability of five-dollar phone calls. Five dollars to talk for half an hour to your cousins in Bangladesh is a bargain.

Although Harris had become familiar with the intricacies of call-sell fraud, Kaiser's hacker case had been his first glimpse of a whole new kind of crime. Harris had been watching, amazed, for months as the evidence accumulated that unknown hackers were climbing around in every sensitive phone company computer that existed. He knew the implications as well as Kaiser. He knew the extent of the harm these hackers were capable of doing. He worried now that he was witnessing the explosion.

In the midafternoon, Harris phones AT&T. He calls the long-distance carrier's operations center in New Jersey. That's where the big maps glow red and top officials are tearing out their hair. Harris explains his concern, then volunteers to help investigate the crash.

A beleaguered AT&T official says, "We'll get back to you."

Don't wait for someone from AT&T to get back to you. It's better to answer the question now, to know as soon as possible if the AT&T crash is related to the hacker case.

Harris immediately dials again. Kaiser answers.

"Tom, we need to know every single number the hackers have dialed in the hours preceding the crash," Harris says.

Of course, Kaiser and Staples already know every single number. They've been staring at every single one so long that they start to feel *hypnotized* by the damn numbers.

The more difficult task is to figure out if any of the numbers the hackers dialed is a clue to what caused the crash. Staples comes down to Kaiser's office, where they print out the list from the electronic DNR. It's easier to read that way. Then they go over the list yet again, this time checking off each phone number as they eliminate it. Here's a call from one hacker to another. Check it off. Here's a call to New York Telephone. Check it off.

By late afternoon, Kaiser and Staples are finished. They are relieved. They pass along the news to the Secret Service.

"There's nothing here that would have caused any particular jeopardy," Staples concludes. At least, not within the last twenty-four hours.

* * *

And by evening, AT&T was taking the blame for the crash, announcing that internal system problems caused the failure. That was another relief, because it meant that nobody seriously believes that the hackers did the damage.

It turned out that the failure could be traced to a routine update of AT&T software. Three weeks earlier, software that controls 114 superswitches nationwide was changed, ironically, to improve the reliability of the system. A minor problem had developed first within the Broadway 51, a switch in Manhattan that was trying to obey the buggy lines of code that had been loaded into it three weeks earlier. The switch shut down to reset, just as it was supposed to whenever it swallows a bug, putting its calls on hold for up to six seconds during the reset process. But in the meantime, a second computer developed a similar problem, and then a third, and then before you knew it, several switches were shutting down to reset, diverting their calls elsewhere to other computers that couldn't handle the load. So the backup switches got caught in the reset loop, too.

By seven P.M., AT&T had figured out how to patch the system to avoid the bugs in the code that set off the reaction. The system recovered.

In Manhattan, at New York Telephone headquarters, Kaiser and Staples recover from their anxieties.

And in Secret Service headquarters, Rick Harris recovers from his worries.

But the events of the day have made them uneasy all the same. What if.

Nearly two weeks later, Mark stands outside a little neighborhood grocery store on Corona Avenue. In the dark and the cold, he hangs out with a group of guys from his block, everybody smoking and talking, interrupting their evening walks from the subway to supper. Mark puffs a mentholated 100, as usual. He feels pretty good; he's no longer fretting over the AT&T crash and no longer worries that hackers caused it. His only real worry, right now, is the cramp in his right foot. He flexes, trying to work it out.

So it catches Mark by surprise when someone in the group, a kid who lives next door to the Abenes, says, "Something strange is going on over at your house."

"What do you mean?" asks Mark.

A bunch of men are going in and out of his house, carrying boxes out to cars. They've been there for a while. No one else has gone in or out. A lot of the neighbors have noticed. It's kind of weird.

"One of them told my dad to move his Jeep," the kid says.

Mark forgets the cramp. He stubs out his cigarette, and heads off toward home, not exactly running, but moving as fast as you can *without* running. It's a one-block walk, but it takes him from one ecosystem to another. He turns the corner, hustling from the tropical-colored storefronts and strings of flashing lights of Corona to the frosty darkness of Alstyne Avenue's row houses. He passes by the place in the street where many summers ago a long-lost teenage gang, the Alstyne Avenue Boys, carved initials in the hot tar to claim their turf. Mark is really nervous now. He always feels stress first in his stomach, and it's churning now. Who could be at his house?

He can see the house now, but Mark doesn't see anyone outside. He's breathing a little hard from the hustle, and he wonders fleetingly if the kid was putting him on. He's filled with a moment of hot relief, but then he's suddenly seized by an even stronger feeling: dread.

Mark skitters up the stoop to the front door with the three colored panes of glass. In one pane, his mother has taped a picture of the Virgin Mary. As soon as the door closes behind him, Mark hears knocking. Someone's out on the stoop. He opens it, and a tall stranger wearing a three-quarter-length tan quilted jacket blocks the entryway. Mark is scared—in fact, he's freaking out. The man says his name is Agent Russo. He says he's from the Secret Service. All Mark can think about is how cheap his jacket looks. It's an odd thought, totally inappropriate to the moment, but it's the one idea that Mark's mind consciously reaches out to grab as a million other urgent images fly past. It's the thought Mark uses to divert himself from the terrible truth of what is happening in his house.

He's being raided.

No, actually it's a hacker named Phiber Optik who is being raided. But wait, no, it's not just Phiber Optik taking the fall. Mark Abene, age seventeen, a junior in high school, avid reader of the book *Gnomes* as well as the circuitry repair tips of Forrest P. Mims III, enthusiastic intern at the Queens Hall of Science, is being raided by the Secret Service.

Mark lets Russo enter the house.

Other agents appear in the living room. Mark figures there must be nearly a dozen of them, all wearing those dark blue windbreakers you see on TV cop shows, jackets that say U.S. SECRET SERVICE on the back. Mark sees the holsters under their jackets. It's an incongruous sight, all these big gruff agents filling the narrow dark space of Mark's mustard-colored living room. Everything looks just the way it did when he left for school this morning, but now it's hostile territory, invaded by an occupying army: navy blue sofa, dining room table swathed in a clear plastic cover, the coffee table books about Dom DeLuise and Marilyn Monroe, his dad's jazz records stacked so neatly, his mother's shiny black boots lined up on an old *Newsday* by the front door.

Soon the agents tell Mark all his computer stuff has been packed up and removed from his bedroom. "The room is so empty you can paint the walls," one agent says.

From somewhere, Mark's mother and father appear. They seem to be floating in the doorway, floating into and out of Mark's vision, as if they were next to the mirror at the Hall of Science.

An agent tells them, "Your son has caused billions of dollars' worth of damage."

The agent is referring to the AT&T crash. Wait a minute! All the agents seem to think Mark is responsible. There's been some mistake! It was all just a software bug! But Mark can't scream it out. He finds himself unable to speak at all.

Everyone sits down at the dining room table, Mark at the head. For some reason, all he can do is focus on each intruder, taking a mental snapshot of each face around the table. One weather-beaten old guy with a gray moustache in an Australian cowboy hat with one side of the brim turned up. Two men from New York Telephone security, including one so old he had to put on glasses to read. Agent Russo, thin with plain brown hair and a bald face. One of them says to Mark, "I'm from Bellcore in Piscataway."

A Secret Service agent says, "Pumpkin Pete says hi." This (false) statement has the desired effect of making Mark think that his friend ratted him out. Mark is starting to feel paranoid. (But can you just be "paranoid" if the thing you're paranoid about is really, truly, absolutely happening—happening right in front of you in the dining room before supper?)

It's all happening really fast. An agent says, "We want to ask you some questions." They would like Mark to come back to Manhattan to headquarters with them, though he is not under arrest.

Mark's parents finally get angry. They say no.

Mark's parents tell the agents to leave. Then they look at Mark. The agents look at Mark. The Bellcore guy looks at Mark. The New York Telephone security guys look at Mark. Everybody wants Mark to speak, to *say something.* Finally, they leave, funneling out into the chill January night.

And Mark has never said a word.

Thirty miles east, a car pulls up to the lighted security booth at the entrance of Polytechnic University on Long Island.

The driver asks for directions to the dormitory.

Five minutes later, Paul Stira hears a knock on the door of Room 4B, the dorm suite he shares with three other students. It is around six o'clock, and the roommates have just finished cooking dinner. Paul opens the door and sees the director of student life standing in the cluttered hallway, looking stricken and flanked by three burly men dressed in business suits. One of them has his jacket off and wears a holster and gun.

Paul is scared, but he's not showing it. His face is a blank. If you went just by his expression, you could just as easily believe Paul was watching a TV rerun as being raided by the Secret Service. Inside, Paul's roiling. But it sure won't help to let the Secret Service know that.

One of the men introduces himself as Special Agent Jeff Gavin of the Secret Service. Gavin looks over Paul's shoulder into the common living room with the brown couch and his gawking roommate. Gavin asks Paul if there's someplace private to talk. Paul, typically, is noncommittal. Gavin finally asks the roommate to leave.

What happens next is in dispute.

The way Paul remembers it, Gavin comes down on him like a tough guy, tells him that his older brother, Tom Stira, has been arrested for credit card fraud that morning and that Tom "told us everything" about Paul's hacking exploits. This is very confusing to

Paul, because as far as he knows, his older brother has never been involved with any kind of fraud, and doesn't know the first thing about computers. In fact, Tom Stira never was accused of any crime.

Then, Gavin waves a manila envelope around, with the figures $70 million and $40 million written on it, and says, "You're responsible for a $110 million lost to the phone company." That would be the AT&T crash again. Doesn't the Secret Service believe AT&T's explanation?

Gavin asks, "Are you Charles Stira?"

"No," says Paul. "I'm Paul."

Gavin remembers a very different version of the scene. He had not heard thing one about the hacker investigation until agents in the city called him in the Melville office at five P.M. They told him that New York City agents have already arrived at Paul's family's house in Cambria Heights with a search warrant, only to find that their quarry and his computer were at college. Fearful that Paul would get wind of the visit before they could drive out there with a new warrant to search Paul's dorm room, the agents called Gavin and asked for assistance. Gavin took notes on a manila envelope, he says, but he later lost the envelope. He doesn't remember what he wrote on the envelope, and doesn't remember saying that Paul was responsible for $110 million in damage. Gavin says he doesn't remember telling Paul his brother had been arrested for credit card fraud.

In one version of this story, you have a bully trying to intimidate a college student. In the other, you have a helpful federal agent introducing himself to Paul. The rest of the events are not in dispute.

Gavin asks Paul for written permission to search his dorm room. He tells Paul that if he doesn't sign, then they'll all have to sit around together in the dorm until a judge signs a search warrant. That could take hours.

Paul asks if he can call his mom. Gavin says he'll go with Paul to the pay phone in the hallway. He stands right next to Paul, even dials the call. Paul feels like a prisoner.

His mom answers. "What have you done?" asks Jean Stira, who has hurried home from her job as a secretary at the Queens District

Attorney's office after learning that Secret Service agents were in her house. Now, agents mill around her as she talks on the phone to Paul.

"I don't know," says Paul. "It must be something with the computer. It must be something through bulletin boards or something they got my name from."

"Cooperate and do what they say," she advises him. In Queens, agents are confiscating all kinds of stuff from the basement. They take Paul's dot matrix printer. They take a spare keyboard. They take his Master modem, and his Volks modem, too. They take a fax machine, and a Western Electric telephone. They take his Commodore 64 programmer's manual. But they don't stop there. They take a box of fourteen cassette tapes, a baseball scorers' sheet with various notes, and personal letters. They take something that they label "Notes to Hack Canadian Ministry of Technology." (The notes about hacking our neighbors to the north are actually notes that Paul jotted in 1985 while watching a fictional television program called "Hide and Seek.")

And now, in his dorm room, Paul signs. Gavin hands him a one-page form titled CONSENT TO SEARCH, and Paul reads it over and scrawls his name at the bottom.

Then Paul watches while one agent takes Polaroid photos and another carries away boxes full of his computer equipment.

Out the door go the last vestiges of Paul's identity as Scorpion: his computer, a hundred-and-fifty floppy disks, more modems, his connection wires.

When they're finished, Gavin tells Paul that some agents from the New York City office will contact him shortly.

In fact, they do, the very next day. They tell Paul that they left behind a printer, a joystick, and a mouse, and would it be okay to remove them?

Paul says OK.

It is nearly midnight, and Eli stands in the lobby of the Secret Service headquarters in the World Trade Center in lower Manhattan. This is a building that in a few years would be the site of the worst terrorist bombing ever on American soil, but tonight the place is empty and quiet as a library.

Eli is alone in the lobby. He waits. He admires a large Plexiglas-covered display of U.S. currency and another that shows all the presidents. Until recently, these were the two major mandates of the Secret Service: protecting the currency from counterfeiters and protecting high-ranking officials. It's the former role that has evolved into fighting credit card fraud and computer crimes.

How is it that of the three MOD boys that the Secret Service raided today, Eli is the one who finds himself standing in this antechamber, waiting to confess?

The scene when he'd arrived home from the movies at around seven o'clock was chaotic. A tall man in a suit had come out of the house with a box. Eli said hi, and the man said, "Who are you?"

"I'm Eli," he said.

"Do you know who I am?" the man said.

"Yeah, you have a badge around your neck."

When Eli opened the door and walked in, a group of men he'd never seen before all yelled in unison, "Eli!" and it scared him.

Eli stood in the dining room, near the doorway to his bedroom, and there saw his mother, Maria, very upset. He greeted her in Spanish. She alternated between crying and digging out family photographs to show the Secret Service agents evidence that the Ladopoulos family had good friends in Costa Rica. One of the Secret Service agents said, "You're kidding. I was just guarding the president of Costa Rica." And Maria pulled from the buffet in the dining room a snapshot of the Costa Rican president himself. Her sister had worked for the Costa Rican Embassy at the United Nations and as a kid Eli liked to go there and try on the headphones and listen to all the different languages.

The way Eli remembers it, the agents told him he was in a lot of trouble and said that if he didn't come downtown with them right now to cooperate, it would look bad later on. The way Eli remembers it, one of the agents asked him if he was a Communist. They asked him what his hacker "alias" was, and Eli told them, "Acid Phreak . . . it's like a pen name."

Now, several hours later, Eli still hasn't had dinner. He's waited for nearly an hour in this lobby, flopping on the couches, standing up again, pacing. Finally, Special Agent Martin Walsh comes out to escort Eli into a conference room.

Walsh, Harris's colleague and the lead agent on this case, looks like what you'd want in a career cop. Neutral. He's a block of a man, square and sturdy, no soft spots. You can try as hard as you want to describe him, but you won't get the color of his hair right. It's not blond, it's not brown, it's not black. It's not straight, it's not thick, it's not thin. His eyes are deep set and earnest.

To Walsh, Eli seems unusually cooperative and eager to give a statement. Walsh thinks Eli is cocky, because Eli keeps boasting about how talented he is on computers. Walsh tells Eli he is not under arrest and is free to go whenever he wants. He also reads him his Miranda rights: Eli has the right to remain silent, anything he says may be used against him.

Walsh says that if Eli cooperates, well then, Walsh will let the U.S. Attorney know. That could help Eli's case.

Eli cooperates.

He writes out a three-page statement:

Having started playing and learning about the telephone network at an early age (since about ten years old) I feel what I know should have been put to good use. Hands-on experience was impossible at age fifteen (I got my first computer then, I only played with the phone from 10–15 yrs old), so I basically started learning from others interested in learning such as I.

Being that the telephone network was easily one of the most technically prominent was probably the reason I chose to learn more about it. I think I should make clear my standpoint on 'hacking' and 'hackers' in general.

Hacking to me is more than just hacking out 'PIN' codes (which I do not do). Any idiot can leave his computer running a software program to scan from them all night. Many consider themselves hackers, they post as many as possible or trade them (on bulletin boards—BBSs).

I particularly do not like this 'hacker's scene' since it misinterprets what hacking really is to me: learning about a system by asking a lot of questions. I knew from the start I would someday work as a computer security consultant or investigator for the telco or gov't agency. What troubled me the most was: how was *I* supposed to come out and say, 'Hey, I'm a hacker. . . . I know this and this about your systems!! I need a job,' without the possibility of being attacked. . . .

So I set out to find out more about the system by reading documentation and by using the systems. Using 'PINS' was only a 'bridge' to come off a tandem trunk in order to gain entry into the system I wanted to learn more about. I pay for *most* of my calls and could have had free service had I wanted it, but that would be going against my primary goal, to know and learn.

Eli stops writing and hands the confession to his inquisitors, who now include a Bellcore investigator as well as Walsh. They stand above Eli, and they look very disapproving.

"I'm done," Eli says.

They look it over and one of them says, "This is like an inaugural speech."

"I don't know what to tell you," Eli says.

One of them leaves for a minute, then returns with a stack of technical manuals. "We already know what you did, so you might as well tell us," he says.

Eli is scared. "Can I call my mom first?"

But then Eli agrees to give it another shot.

Walsh says to Eli, "You know what a switch is, right? Put that in there."

They discuss MIZAR, COSMOS, LMOS, and Eli helpfully writes out four more pages of detailed explanation that define these acronyms and the extent to which he explored those computers. Eli notes that of all the phone company's switches, DMS 100s "are my favorites."

I set up forwarding lines in order to reach people that were far for me to be called 'locally,' thereby *I wouldn't have to use another person's* 'PIN.'

. . . My intentions were not criminal in my eyes. My only intent was to learn, and in the back of my mind, I knew I wanted to work for a telco, whether RBOC or interlata [long-distance] carrier, I did not care. This is what I know best and my career was spelled out. If I had had on-hands experience or had been working *for* telco, I wouldn't have done what I did. I saw no 'channel' I could use to get a job with them (telephone co.), so I applied with a management company in hopes of getting a computer-related job.

. . . I was a 'pioneer' in my eyes, since me and my friends were more advanced than the majority of the 'hacker population.' I learned everything on my own and from friends, but had no criminal intent. I feel it is a waste of talent when the 'hackers' are viewed, generally, as evil mischievous people.

. . . I was looked up upon by most of the 'lesser' hackers and I would guide some and help with programming (*not* to enter systems, I *never* posted a dialup or password to telco systems on *any* system).

Eli goes on to write about all the times he called long-distance carriers and New York Telephone looking for a job, and notes that no one ever returned his calls.

He writes that he never meant to do any harm, "which I feel sorry for if I have. This is not B.S., or a cop-out. . . . I just think my side should be told, and I am happy to cooperate."

Walsh looks it over, then says, "Okay, this is better."

Walsh shows Eli a sheet of paper on which is written the electronic message that flashed across the screen when The Learning Link crashed the previous fall: "HAHAHAHA . . . Happy Thanksgiving you turkeys, from all of us at MOD."

Eli writes across the sheet of paper: "This is a true copy of what I put in the computer." He signs his name.

Then Walsh says, "Just write here, at the bottom, that you did this of your own free will."

Eli signs his name again.

The session is over, and now Walsh brings Eli into a small room in the back of the office where there is a coffee machine and a telephone on the wall. Walsh gets some coffee, and Eli calls his mother to tell her he is okay and will be coming home soon.

Eli is so confused. He just wants to get out of here.

NINE

John Lee dials "0." The hacker leans against a bank of pay phones in the atrium of the soaring Citicorp Building in midtown Manhattan. He's much too cool to check out his reflection in the plate-glass window of the Italian restaurant at the edge of the open space.

All around him, teenagers peck at phones like sparrows at a birdfeeder, dialing and hanging up, trying random phone numbers. About fifty teenagers are hanging out, swapping information on how to break into computer systems, forming energetic clusters. Kids dial a number that someone said would connect to a New York Telephone switch. The number worked last week.

The hackers had materialized out of nowhere around six o'clock, just when the office drones fled the Citicorp building for the subway. The kids came here once a month, on the first Friday of the month. No matter the season, they wear beat-up fatigue coats, baggy jeans, clunky-soled shoes, thick-thick black belts with square metal buckles that weigh five pounds. They sport peach fuzz moustaches and slicked-back buzz cuts. You can't miss them. The gathering is their "meeting," and it's organized by the semi-official hacker house organ, a quarterly magazine called *2600* that's published out of Long Island. The magazine's name is an allusion to history: back in the Dark Ages of the 1960s, a 2600 hertz tone was used to control all the phone company's trunk lines. Today, the tone is obsolete, but the magazine still pays homage to the days when a street-smart hacker named Captain Crunch figured out that blowing a freebie whistle from a cereal box produced the precise tone necessary to make free phone calls. The articles in *2600* tend toward telco technicalities. Every issue sports a photo spread of pay phones around the world, hacker pinups: a pay phone kiosk in Belgium, airport pay phone in Rome, London phone box. The magazine is eclectic, and so are the meetings held in its name. There's no agenda at a *2600* meeting, no call to order, no parliamentary procedure. It's much more informal than that. The editor of *2600* is Eric Corley, and you can see him today cutting through the crowd, long black curls flying, as he passes out copies of the latest issue of the magazine. At thirty, Corley is the dean of the Citicorp crowd, a college kid who never left campus, a radical who never mellowed with age. He has a weekly radio show in Manhattan (tune in on Wednesday nights to hear him on such esoteric topics as how New York Telephone could trace kidnappers' ransom calls more efficiently), and for all intents and purposes, in 1990 Corley represented hackerdom to the rest of the world. He was the unofficial spokesman. Or at least he was until recently. The Secret Service raids had changed some things.

In fact, in the weeks since word of the raids first spread, the *2600* meetings have become much more of a crowd scene. This afternoon, some hackers have traveled from as far away as Syracuse to make the meeting. A few kids cluster by the potted plants, planning an after-hours dumpster dive. "We got thirty garbage bags of printouts from the Bank of Tokyo a few months ago. Found five passwords," says a chubby high school student wrapped in an army coat.

It's quite a scene, and no serious hacker in the New York City area would miss it. Most of the MOD boys are here; they come every month. One time a dozen MOD boys posed for a group photo. Red Knight, an Indian kid with a penchant for collecting telco manuals, was wearing a sweatshirt that said "MOD" in big white letters. He stood in the middle of the group, and they all draped their arms around one another. Solidarity.

Today, the only MOD member missing is Paul. He's away at college, and he might not be here even if he were home. The raid really scared him, and he's vowed not to hack anymore. Eli and Mark see things differently. They've gone one-on-one with the government, and are happy to supply the details to other hackers.

Paul has been feeling distance growing between himself and his friends, because he's not comfortable with how they behave. He doesn't like to boast. The three of them don't argue about it, because that's not Paul's way. He doesn't confront them; he just withdraws subtly. Paul has never been an easy talker. The joke was that you had to know him for months before you ever got to hear his voice. He'd sit in the background of any social gathering and just listen. Eli and Mark loved to talk, but Paul might not say a word for days. It made some people uneasy, but the fact is that it made Paul blush even to think about talking about himself.

Eli and Mark stand at the center of a gaggle of fascinated acolytes near the payphones. How does it feel to be raided by the Secret Service? Hey, were your parents freaked?

Eli and Mark have become a walking enlistment poster for hacking. MOD WANTS YOU—TO THWART UNCLE SAM. The fear the two had felt upon walking into their houses and confronting a bunch of armed federal agents has faded; it's been replaced by a certain complacency. They'd been raided, and here they are to tell the story. Never been arrested. Never heard back from the Secret Service in the few weeks since the January 24th raids.

Although they lost their computers in the raids, MOD was as alive as ever. Eli has seen to that. He's even gotten back to "The History of MOD" and updated it. The phile was making the rounds among MOD and its closest associates:

MoDm0dMoDm0dMoDm0dMoDm0dMoDm0dMoD0dMoD
^^

[The History of MOD]
BOOK THREE: A Kick in the Groin

Well, suffice it to say, the fun couldn't last forever. . . . the Secret Service visited the homes of Acid Phreak, Phiber Optik and Scorpion. . . .

Days later they had gone to meet The Wing, which wasn't able to talk for too long since he was too busy. He had been anticipating this little visit for a while, though. His dad didn't exactly like the idea of their presence and kicked their lack-of-a-warrant asses out before they got a chance to put to use their years of interrogation techniques

classes. Seems they think he showed his teacher a credit report or
something. . . .

The group's popularity soared in such a short period of time, but
many hackers disagreed with the MOD style much in the same way
Phiber Optik had enjoyed humiliating those "in the know" publically
[sic]. . . .

Eli also started a chapter called "MOD Policy," in which he
articulates the group's code of ethics and what he calls "DA
RULEZ."

MOD generally frowns upon mischievous stunts that lead to abuse
of a system, damaging of a system or any type of anarchy for no
apparent reason other than to "be c00l."

When Paul saw that paragraph, much later, he remembered The
Learning Link computer crash. He was amazed at the obvious spin
control his friend was putting on the MOD gang's exploits. He felt
the same way when he read that MOD no longer stood for Masters of
Disaster. Now they were the Masters of Deception, according to Eli.
Sounds less threatening and certainly, less destructive.

The Masters of Deception. That was a good one. Who was
deceiving who?

Of course, there was plenty of reason for spin control in those
days. In the wake of the raids and the *Harper's* forum, the media had
discovered MOD in a big way. In fact, the whole hacking scene was
making the news. It was a heady feeling, especially for Eli. Six
months ago, he was in the background, just a kid hanging out with
friends who knew a lot more about the technical aspects of hacking
than he did. Maybe he even felt a little insecure about it. But the
Harper's forum had given him stature. He was a national celebrity
now. He was Acid Phreak. Now, everybody wanted to hear what Eli
had to say.

Tonight, in fact, two reporters from *Newsday* mingle among the
adolescents in the atrium's open expanse of ferns and chrome rails

and fresh-juice joints. The *Village Voice* is doing a really huge piece about MOD, and Eli has taken that reporter under his wing. *Esquire* wants to do a primer on how to hack. Geraldo's people are sniffing around, and a screenwriter is even talking to Mark about a movie.

Everything is a media event these days. And what the media want to see is some real live hacking. Break into things, they plead. Show us how to social engineer, they beg. Log in to the phone company, create an unbillable phone number, smile for the camera. To help his friends navigate the unknown, Eli writes in "The History of MOD":

> Mention should be made to all other members of a media event concerning the group in any way. (Don't hog the show) The media is a VERY tricky thing to deal with. They are looking for sensationalism.

But wait: Aren't Eli and Mark the subject of a major federal investigation? Wasn't thousands of dollars' worth of their computer equipment just confiscated?

It would make sense to lay low. On the other hand, all the attention is a little overwhelming and it really doesn't *feel* like they're in big trouble. Instead, it feels like they'd just got discovered. John Perry Barlow has even recovered sufficiently from the mesmerizing appearance of his credit history on his computer screen to travel to New York, where he'd met Mark and Eli face-to-face. Well, actually, he'd been in town on some other business. An NBC producer wanted Barlow to defend the use of LSD on some TV news show segment, since Barlow had been, after all, the one who'd gotten the Grateful Dead to come to Millbrook in 1967 to hang out with Timothy Leary.

The NBC producer happened to be dating the *Harper's* editor who had put together the hacker forum. So somehow, Barlow found himself at dinner with the editor, the producer, and the hackers. They all met at Shun Lee, a Manhattan restaurant known not only for its delicious Chinese menu but also for its sixties' sophisticate decor (James Bond meets red-eyed dragons). Mark and Eli were immaculately dressed and on their best behavior when they slid into one of the long low booths that skirt Shun Lee's perimeter. Bravely, they ate jellyfish.

Barlow thought he knew exactly who these boys were. He recognized an earlier incarnation of himself. He ordered bourbon and water and slipped the drinks to (underage) Mark. The boys, although obviously spooked by the recent raids, weren't about to back off from their hacking activities. They said it was a matter of honor.

Barlow couldn't help but think of beavers.

He used to go to war with beavers on the ranch. They'd get into the irrigation ditches and build dams. As soon as Barlow would blow up one dam, they'd build another, twice as big and a lot uglier. It put those beavers under a lot of pressure to build those dams under those conditions. But they couldn't help it.

And here was Mark, a beaver in Barlow's eyes. Mark's turf was blown up by the feds and he was still coming back for more. The raid only validated his experience, only made hacking more worth the effort. Here was a boy who had lived the better part of the first decades of his life in obscurity, devoted to his obsessive pursuits, who was unknown outside of a tiny circle of enthusiasts. And then suddenly one day he woke up and was famous. He was alive, he was important, he was brilliant. He knew all sorts of things about forbidden computer systems, and he no longer was just a kid filling up the long hours between "Star Trek" reruns. For the first time in Mark's life, he was *somebody*—he was a public enemy, worthy of the attention of the government of the United States at the highest levels. Not to mention the attention of *Harper's* magazine, *Esquire* magazine, "Good Morning America," and *The Village Voice.*

Soon after that dinner, Barlow returned to Pinedale and felt like a furtive beaver himself. It was a coincidence, maybe, that the FBI cornered Barlow, too, but it was an event that had far-reaching consequences.

An FBI agent came to call at the ranch, asking Barlow an afternoon's worth of questions. Someone had stolen the proprietary code embedded in the microprocessors of Apple's Macintosh computers. The code made Macs unique, and now someone was anonymously mailing it around on floppy disks (even Barlow had gotten one in the mail). Did Barlow mind talking about the theft?

How the FBI had gotten Barlow's name was anybody's guess. But the whole experience made Barlow realize how profoundly ignorant the entire federal government was about computers, civil liberties, and the electronic frontier.

As he would later write about FBI agent Baxter, in a ground-breaking manifesto he called "Crime and Puzzlement":

I . . . found in his struggles a framework for understanding a series of recent Secret Service raids on some young hackers I'd met in a *Harper's* magazine forum on computers and freedom. And it occurred to me that this might be the beginning of a great paroxysm of governmental confusion during which everyone's liberties would become at risk.

Barlow posted his writings on the WELL, where they were read by Lotus 1-2-3 creator and zen millionaire Mitch Kapor.

Kapor, who lives in Boston, was struck by Barlow's description of the FBI agent's visit to the ranch because Kapor coincidentally had been visited by an East Coast version of Agent Baxter. Kapor was also unnerved at the direction the world seemed to be taking. Obsessive about computers himself—in his youth Kapor had driven from Boston to New Hampshire to avoid the sales tax that stood between him and his first Apple II—Kapor empathized with the plight of Phiber Optik, Acid Phreak, and Scorpion.

One day, Kapor called Barlow from his private jet and asked if he could detour from a scheduled business trip to visit Pinedale. There was a local airstrip, so within hours, the two men were hatching a plan for a new group, a national organization that would fight vigorously to protect civil liberties in cyberspace. The group's name would be the Electronic Frontier Foundation.

Kapor and Barlow decided that Mark and Eli needed some expert legal advice. With Kapor signing the checks, a meeting was arranged for Mark, Eli, and Paul with a lawyer at the sprawling Manhattan offices of the late Leonard Boudin, who at the time of his death months earlier was among the best known civil liberties lawyers in the country. Boudin had been the man who protected Cuba's U.S. interests for decades, and who represented Teamster leader Jimmy Hoffa after his prison release.

The boys' meeting there in early 1990 was heady stuff. Subsequently, Kapor paid for Mark and Eli to fly to Boston to tell him their story in person. Mark got queasy on the plane; it was his first flight. No charges had been filed, no lawyers formally retained to defend the MOD boys. But the promise of protection seemed implicit. And protection would be welcome. Who knew how far this nonsense would go otherwise? MOD was not the only hackers' group under attack. The government had just raided three LOD hackers in Atlanta. A sting aimed at various hackers across the nation, called Operation Sundevil, was under way and, to the adults at least, the government was overreaching its authority. One man, an *adult* from Texas named Steve Jackson, was threatened in 1990 with the loss of his business, because federal agents had seized his company's computer system. It seemed they'd been after an LOD guy named The Mentor, an employee at Jackson's company, which created role-playing games in the spirit of Dungeons & Dragons. Jackson at that time was working on a fantasy game about computer hackers, which the government apparently believed was a handbook for computer crime. Of course, the government's true agenda was anyone's guess. The government didn't have to tell you what it was up to. One day agents could show up at your door, hand you a warrant, and take your stuff. Out the door would go the precious computer systems of Steve Jackson Games.

And the maddening thing of it was, no one was getting arrested; the government seized this stuff, then sat on it. Jackson would later be vindicated, but that was of no comfort *now*. The situation resembled a scene from George Orwell's *1984*.

The times are perilous, but at least hackers are now getting the attention they deserve. With all the excitement, MOD's ranks are growing. One recently inducted member is The Plague, who studies programming at a Manhattan college. The Plague is the scion of Russian émigré parents. He has a light brown pony tail and high cheekbones, a good-looking kid who looks like Ilya Kuriakan in the "Man from U.N.C.L.E." The Plague is paranoid, and at today's gathering, he keeps close watch on anyone he suspects might be a government informant. "Follow him," The Plague hisses to a friend as a hacker named Long John Silver surreptitiously slips up the stairs to Third Avenue.

There's The Seeker, the "electronics" expert who loves to build all manner of devices to cheat the phone company. Mark likes to stay overnight at his Manhattan apartment.

John Lee is in MOD now, too. So is his friend Julio Fernandez, a kid from the Bronx who uses the handle Outlaw. Julio's in the group basically because he's John's friend. John's induction into MOD was only natural, because of his intense thirst for computer conquests; all he wants to do is penetrate new systems. He's learned so much about hacking and cracking in the one year since he got his modem that he's caught up with the rest of the MOD pack. He'll do anything to get into a new computer. He breaks into a network, then turns over the information to the other MOD members. He's the scout; they chart the territory.

On the pay phone, John connects to an operator. "Hey, I'm up on a pole here," he says, using his deepest, most authoritative, most adult voice. It's Supernigger's signature line and has become standard patter; they all use it now. If only the operator could see John, a tall, twenty-year-old black kid in a white T-shirt and khaki pants so baggy they could hold a friend. He doesn't look at all the part he's playing—a white, middle-aged, tool-belted lineman doing a service check. But he sounds the part. And maybe that's enough. Just maybe the operator will fall for his smooth line of technobabble and give him an open line.

"Yeah, I need—damn."

Disconnected.

John hangs up.

The hackers are all watching, but not watching, on this drizzly not-quite-spring afternoon. It would be lame to pay too much attention. On the other hand, John's stunt is the most interesting thing that has happened today.

John has been hanging with Eli and Mark electronically ever since the raids. But mostly, he hangs with Julio, who looks enough like Eli to be his little brother. He has black hair cut to a fade, intelligent brown eyes, an eager, teach-me attitude. He wants to know as much as Mark and the others. Julio's a good dude, a little young, but he has a nice swagger, a Bronx-tough way of talking when he needs to. He doesn't let anyone push him around. Julio's a little cowed by Mark, though. Most people shrivel under Mark's technical cross-examinations. Julio doesn't shrivel. But sometimes he keeps his distance.

There's also a serious core group of hackers who come to the *2600* meetings with hopes of being anointed by MOD. They want to be accepted, but for some reason or another (maybe the MOD boys can *smell* how bad they want in, and that odor is repellent) the wannabes stay on the fringes. They don't know enough. Or they act like they know too much. Whatever. The wannabes offer Mark and

Eli tidbits of information. They boast about their own exploits. They tag along to Around the Clock, a diner in Greenwich Village where the hackers go to eat after the meeting breaks up. But they won't get into MOD. Ever.

Alfredo DeLaFe suffers that fate.

There's one of him in every single classroom in America. He's the boy who's smart and eager, who wants to be accepted but can't help being a little bit of a know-it-all at just the wrong moment. Has to trumpet the fact that he knows the capital of Mozambique. Of course, the teacher likes him, and sure can't figure out why the other kids pick on him. There's the problem right there.

Alfredo was cherubic, and that's never helped. He was only fifteen, and that sure didn't help, either. He had black ringlets, which probably endeared him to his aunts, but definitely did not broadcast a cool image. He tried to imitate the way the others dressed, in heavy hiking boots and black T-shirts, but just when he thought he'd pulled it off, someone noticed the gold ring on his right pinky.

Alfredo used to run his own electronic bulletin board from the second-floor walkup where he lived, half a block from Central Park. Ten blocks south, and he'd have a tony address. As it was, even up in the mid-Nineties, he was not far from a really cool two-story McDonald's on Columbus Avenue. A space age McDonald's, with kind of a Star Trek theme. That was one thing in his favor, as far as the MOD boys were concerned. That was a reason to visit him. You had to eat.

Alfredo called his board Crime Scene, a pretty boastful name for a warez board that didn't attract any heavyweight hacker types. Even worse, he was just asking for trouble with his I'm-so-cool handle: Renegade Hacker.

The first contact Alfredo had with any of the MOD boys, well before today's *2600* meeting, was when some of them called up to harass him. Alfredo didn't know who the callers were, he says it could have been either John or Julio or Zod on the phone, could have been all of them, for all he knows.

They said, "Give us access to your board or suffer the consequences."

Alfredo hung up, but they called back on his other private line immediately. Then they called again on his mom's phone line. He was impressed that they had all his numbers.

A week later, Long John Silver phoned Alfredo and filled him in on the whole MOD thing. Who they were. What they did. Why he should be friendly. He got the picture, and now Alfredo really wanted to be in MOD.

He even set up an exclusive sub-network on his bulletin board so the MOD guys could log in and hang out, like sparrows at a bird feeder. They called a couple of times, but then Alfredo's mom made him take down the board because his grades were slipping. Thanks, Mom.

Of everybody in MOD, Julio was the most cordial to Alfredo. Julio called him on the phone sometimes, and he called him "Alf." He even came over to Alfredo's spotless apartment with the sculpture of the Madonna and Child looming over the dining room table and the shiny linoleum floor that the chihuahua is *not* allowed to walk on. The yappy dog stays in the hall, behind a toddler guard-gate. You could hear the dog's toenails clacking back there all day.

Sometimes Julio brought John over to Alfredo's. But it was a lot different from the scene in Eli's bedroom. It was no meeting of the minds. For one thing, Julio and John didn't really like Alfredo, at least that's what Alfredo thought from how they acted. One time after they left, Alfredo was missing a modem, worth maybe fifty bucks, and a beeper as well. He suspected that John stole the stuff, but John remembers the situation differently. He says Alfredo wanted him to have the modem. Also a pocketful of change.

It really burned Alfredo, if you must know. You let some guys into your house, you think they're your friends, and then they pull something like that. Alfredo might be guilty of bad judgment, but he was no geek.

He invited Julio and John over again, pretended to forget all about the modem and beeper. He played the lamer supplicant: Show me some stuff on the computer. Pretty please. Julio and John were only too happy. (If they hack into stuff from Alfredo's building, how could the feds ever trace the intrusions back to them? Slick.) They would take him on a tour, a hacker's tour of Fun Sites. Julio would do the typing.

They go to New York University, calling a campus computer system that lets them dial out through another modem. From that point on, whatever they do is very hard to trace, and even better, free. To them at least. They take Alfredo to the Internet (big deal) where they show him a whole collection of special-interest bulletin boards known as USENET. They browse one board dedicated to discussing telecommunications. It's just like a hack/phreak board, except it's for adults. No one uses a handle. The users sign their real names! What fools.

Show me something better, says Alfredo.

They log in to TRW. Easy. Nothing to it. Who do you want to look up?

Geraldo.

Of all the people in the world whose confidential credit histories are online, whose would you like to pull? Geraldo Rivera's. The dude's mean, in your face, and right to the point. Just like them. He's Latino and he's from the city.

A credit record spews out:

```
JK2N RIVERA G.0 < 10000
9 10: 10: 04 CU06 502 RIVERA TNJ1
A-P11 PEOPLES WESTCHESTER BK 1008880 0
```

Could it really be Geraldo's? Cool. The talk show host's credit history goes on for pages. American Express, a student loan from Florida, a credit inquiry from World Book, a Citibank preferred VISA card, a satisfied loan to Manufacturer's Hanover Trust, and a mortgage through Citicorp. Of course, they don't know for sure if it is the right Geraldo Rivera. But it's the only one who comes up in a classy neighborhood in the metropolitan area.

Eventually they tire of Geraldo and his secrets and move on, to the Information America Network. It's an online service that has all sorts of interesting personal data about you and me. That's how Information America makes a living. You have a civil judgment against you and it goes into the network. The kind of house you live in, the names of the other people who live there, even your neighbors.

```
INFORMATION AMERICA NETWORK
P E O P L E F I N D E R
```

```
(Copyright Information America Inc.)
Client Billing Code: 7689
Who do you want to look up?
```

But before anyone else can answer, John decides:

```
Last Name:duke
First Name:david
Enter XX=state abbr MU=Multiple state (3)
US=Nationwide:la
Would you like to narrow your search to a
specific city? (Y/N):n
Searching . . .
```

The anticipation is palpable. What would you say to David Duke if you could call him on the blower? What would you say to the redneck Lousiana legislator, the racist who openly admits to being a Ku Klux Klan boss? Within a few seconds, Duke's address, birth date, length of residence appear. Best of all, there's the man's telephone number.

John dials. A man answers. He sounds a little old, but what the hell. "Yo," says John, slowly, dragging it out to three syllables. He sure doesn't sound like a white telephone lineman now. "I wanted to talk to you about your daughter. . . ."

And on and on. It is a wonderful night for Alfredo because, wouldn't you know it, he's been recording the whole session. He records it on his computer, secretly, so that John and Julio don't even know. Every keystroke, every line of text that's printed out on the screen is secretly buffered in a file on Alfredo's computer.

After they go home, he prints it out and admires the captured session. It might be useful some day.

Alfredo adds it to his collection of confidential documents. He's also managed to procure a printout of the updated "History of MOD." That technically may be a confidential MOD phile, but hey, it's not so hard to get hold of, if you know the right people.

John and Julio think they're so smart, dissing Alfredo. But Alfredo knows a lot of the right people. He's friendly with a lot of guys, guys who might be just as fed up with the MOD boys' posturing and crowing as Alfredo is.

Alfredo knows some of the guys in LOD, for instance. Well, he sort of knows them, at least well enough to type greetings when they're all logged in to the same chat system or something. Alfredo knows Chris Goggans himself, and knows all about the feud between Mark Abene and Chris.

If Alfredo can't get into MOD, he wouldn't mind getting into LOD.

As far as Alfredo knows, the LOD boys haven't seen a copy of "The History of MOD" yet.

But they sure would be interested.

TEN

Downstairs, the front door opens.

It is a familiar sound. Although Chris Goggans is asleep in the second-floor bedroom of his off-campus townhouse in Austin, he wakes to register the click of the latch, and to wonder. The door shouldn't be opening at 6 A.M. His roommate isn't even awake yet.

There is no time for the wonder to turn to worry, no time at all it seems, because suddenly, Chris hears a gruff voice shout up the stairwell.

"Federal agents! Search warrant!"

Chris, bewildered in his boxers, wakes to see an agent in his room. Chris sits up in his bed and sees all of them, six men wearing badges and carrying guns in holsters, spilling into the house. If you had asked him yesterday, he would have said no way could that small entryway accommodate six grown men at once.

From the next room, Goggans's roommate awakens, and calls out, "Chris, you better wear socks! They make you take off your socks."

This visit was no surprise. Both roommates know the agents have come to see Chris. The premonitions weren't exactly psychic. Three hackers in New York were busted in January. Three hackers in Atlanta were busted in February. If that weren't enough to give him a prickly feeling between the shoulder blades, Chris recently learned that his student records had been subpoenaed by a federal prosecutor. In fact, for weeks the University of Texas student has been semi-expecting just such a near-dawn arrival, expecting the government to come for the computers.

But Chris will fight them every inch of the way. Not for nothing has he chosen the hacker handle Erik Bloodaxe. His namesake, a brave Viking warrior with bizarre magical powers, captivated Chris when he read about him in a sixth-grade history book. Erik Bloodaxe is a persona worthy of membership in the Legion of Doom. The original Erik Bloodaxe might have slain these gun-wielding interlopers with a single swipe of his massive steel sword, glinting righteously in the sun. Chris is armed with a more modern weapon: smart-ass college student sarcasm.

On Chris's dresser, laid out plain as anything (to be more obvious, it would need a big red arrow dangling from the ceiling, emblazoned with the words LOOK HERE) is a brochure on how to become a Secret Service agent. The raiders whisk it up, turn it over to their leader.

"So you're thinking about joining up, heh?"

"I think I could help you guys out, yes," Chris says.

Thus is the bluster of Erik Bloodaxe translated for the late-twentieth-century ear.

The federal agents allow Chris to dress. He pulls on a pair of jeans, then goes to sit in the kitchen downstairs. He hears them overhead as they walk around the room where he so recently slept. He hears them pack up his equipment.

One of them rustles through papers and finds a copy of a song parody Chris wrote. It's called "The End of the Internet."

"Is this some kind of a threat?" an agent asks him.

"No, just a song," says Chris.

The six-man team is also curious about an arcade-size Pac Man machine. They suspect Chris stole it; he says he bought the game from a friend. They phone his friend at seven A.M. to verify the story. Also snapped up are a bag of stereo cables—wires, hmm—and printouts from an electronic magazine called *Phrack*. The agents dig through cereal boxes, flour containers, detergent, apparently looking for drugs. Blue suds foam up.

They leave the house carrying boxes. Chris walks them to the car, then watches them drive off into the morning. He wonders if he will be arrested. When? He wonders if he will go to jail. For how long? Will it be a big case? Will he be sacrificed on the altar of some prosecutor's career?

Pity Chris. The Viking Erik Bloodaxe could have relieved his anxieties by slaughtering a few goats, at least. But Chris is left at the curb, left to turn and walk back into his townhouse. He closes the door behind him. He hears the familiar muted click.

Again, no one got arrested. And maybe because of that, the whole thing took on surreal overtones during the next few days and weeks. Maybe this was some kind of Big Brother bad dream. Maybe it would all go away. Maybe it was just the government's way of harassing a bunch of kids—you know, shut them down by expropriating their computers.

That approach didn't work with Chris, who was pugnaciously dialing underground bulletin boards again almost as soon as the federal agents disappeared from his life.

There was a new, super-elite board that he had to check out.

It was called Fifth Amendment.

The board ran on an IBM computer in a private home in Houston. What was it that would start the buzz? Why did some boards spring from nowhere and overnight build a reputation for being the new, hot lure, the BBS with the phrankest philes, the baddest place to post?

Fifth Amendment was suddenly a sensation, no less so than a Prohibition-era speakeasy. Hundreds of hackers knew there was a bright, warm place behind the locked door, but only a dozen or so were privy to the password that would get them beyond the bouncer. What made Fifth Amendment so alluring was the crowd inside. The patrons were a collection of the world's best hackers, or so the huddled masses outside whispered to one another.

Word got around. Fifth Amendment boasted an international clientele. Dutch hackers (the *Dutch*) were said to be on the board. So were the guys from the Chaos Computer Club. A man named Empty Promise, an ex-Navy cryptography expert, logs in from time to time to discuss the intricacies of the phone company's packet-switching networks. Some of LOD's rock-solid, original members were on the board. Here, on a personal computer on a desktop in a bedroom in Houston, the hacker underground's elite consorted with one another.

The board's operators, a hacker named Micron (who owned the actual IBM hardware) and his friend Scott Chasin, alias Doc Holliday, kept the list of users very small, very private. In this climate of uncertain civil liberties and Operation Sundevil, where the government cracked down on first one hacker and then another without warning or obvious patterns, a lot of the best hackers had put their modems on a back shelf in their closets. Micron and Scott wanted to bring them back from the dead by providing a safe haven for the free interchange of ideas—and whatever. By being so exclusive, Fifth Amendment would protect its users from incriminating themselves. Hence the name.

Chris logged in, using an authorized user name that came directly from the system operators. The password was changed often, disseminated directly from the system operators to each authorized user.

Chris would read the philes, chat with the other users, correspond for a while with Scott Chasin. Chris and Scott had an electronic friendship that dated back to the mid-1980s, back to the days when they accidentally bumped into each other in the Great Plains of cyberspace. They both were logged in to a midwestern-based underground bulletin board called World of Kryton. Kryton was run by a Milwaukee hacker who boasted membership in the elite gang that called itself the 414s (named after Milwaukee's area code). The gang's fame derived from the fact that members had figured out back in 1983 how to hack into the computer system of Memorial Sloan Kettering Cancer Center in Manhattan. The 414s had messed around with some files for a while, but the main effect of their actions had been to get arrested.

Back in the days of Kryton, or maybe soon after, Scott and Chris discovered they had a lot in common. They were close in age, babies conceived on the cusp of the 1970s, both born and raised in the secure style afforded the white middle class in Texas.

Both Scott and Chris grew up in a place where their status in society was assured by who they were and where they came from. Maybe, at some level, that was the comfortable quality they recognized in each other when they first started corresponding on Kryton.

Or maybe the foundation of their friendship was built on similar childhood journeys of electronic discovery. Both had been intoxicated while still in grade school by the blink-and-whir of a desktop machine. Both endured their parents' divorces. Each had been the third party in computer love triangles by age ten, falling for machinery already claimed by another. There was something illicit about it, which made each keyboard stroke mean more somehow, with Chris hunkered down in front of an old Apple that belonged to a friend's father, and Scott gingerly stealing processing time off a PC that had been a gift to his older sister.

Whatever the source of their affinity, by the time they cavorted on Fifth Amendment, Chris and Scott were unquestionably allies.

The two friends spoke on the phone a lot, too. They did conference bridges. Bridges were the best.

Chris and Scott were part of the mad rush of excited hackers who all glommed on the same open phone line, crowding into a conversation like it was a rush-hour subway. The phone line in question belonged to—Well, let's just say it was temporarily liberated from the phone company, allowing anarchic hackers to engage in huge transcontinental conference calls that bridge across this city and through that state as one kid after another gets on the line. If you were on the line, and you had three-way calling on your phone, you could invite a friend to join the conference call, too. Simply hit the flash button that disconnects a call, then call your friend, then flash again. Your friend is three-wayed in now. And if *he* has three-way calling, he can recruit yet another caller to the conference. These daisy chains lasted for hours, for days, for marathon amounts of time that adults couldn't even imagine. There was so much to say.

Bridges were a great way to get acquainted. You could take a tour of the world on a bridge, talk to one hacker in Holland at the same time you converse with somebody in New York City. In fact, a couple of mysterious New York newcomers named Corrupt and Outlaw brushed up against Texas kids pretty often during conferences. Chris and Scott had never actually met the New York boys, but they'd heard of them. Vaguely. They'd heard that Corrupt and Outlaw came from the "inner-city ghetto," but they seemed to know their stuff.

One night on a bridge about five or six hackers—all kids from Texas, you understand—are hanging out on the line. What were they talking about? Random stuff. Chris says he wasn't on. Scott is on. Suddenly, another voice calls in to the conference, joins the group in mid-sentence. The unknown newcomer does not have an accent common to these parts.

"Yo, dis is Dope Fiend from MOD," the newcomer says in distinctly non-white, non-middle-class, *non-Texan* inflection.

One of the Texans (who knows who) takes umbrage.

"Get that nigger off the line!"

The newcomer is silent.

In fact, the whole conference bridge is suddenly silent, all the chattering boys brought up hard and cold against the implacable feel of such a word. You might as well have slapped their faces. Interminable seconds pass. Who wants to fill that void?

Then the newcomer speaks with a different accent, and the words he says to the white boys from Texas are these: "Hi. This is Corrupt."

And that's how it happened. It was as simple as uttering one ugly word. The racial epithet instantaneously moved northward over hundreds of miles of cable, ringing in the ear of John Lee, who sits at his Commie 64 in his Brooklyn bedroom way at the other end of the line.

That word hit John like a billyclub. Here he was, calling up probably some of the same guys he's chatted with who knows how many times, and the class clown decided to tease them a little. He's got a million voices, you know that, from middle-aged phone lineman to street-corner pimp. He never figured that his joke would elicit such a response.

"Get that nigger off the line!"

Who yelled it? It's really immaterial at this point, because John was going to make them cry.

Nothing would ever be the same again. Not for Chris and Scott, not for the boys from MOD, not for the loose-knit community that made up the hacker underground.

With that one word, war had been declared.

You don't survive on the street by allowing white boys to call you "nigger."

Of course, John didn't say any of this to them. Not then. In fact, they all chatted amiably for a while, long enough for Scott to decide that John really knew his stuff. A few days later, after learning more about John and Julio—and MOD—Scott offered John an account on Fifth Amendment.

You've got a good reputation as somebody who knows a lot about VAXes, Scott said to John. We could use somebody like you down here in Houston.

Thanks a lot, John said.

Just one thing, Scott warned. "We'll give you an account, but don't give it to your buddy Mark up there, or your buddy Acid Phreak." In the wake of the *Harper's* forum and the Secret Service raids, Mark and Eli had become notorious, and Eli in particular was gaining a reputation for being a media hound.

Sure, said John. Right. Whatever you say, dude.

A few days went by, and John logged in to Fifth Amendment, calling himself Corrupt. He read the philes, he typed messages, he abided by Scott's injunction that he shouldn't try to copy anything that was on the board.

Then one day, two new users showed up on the board. They called themselves Broken Leg and Flaming Carrot. They were really obnoxious whenever they logged in. Nobody knew who these guys were. John said he didn't know, either.

John ventured back into conference bridges, bringing along his friend, Julio. The Texas boys were quite civilized to them, real friendly in fact, John thought.

One night, John and Julio called into a bridge and a bunch of people were on the line. John couldn't tell who was talking and who was listening. But he could hear the way the Texans referred to him and Julio—the niggers and spics from New York.

John and Julio stayed quiet on the line. They didn't give themselves away.

This time, John wasn't as shocked.

But Julio is mad, madder than John had ever seen him. They don't know who specifically has been slurring them; they don't recognize a voice, but they know from the accent that it's a southerner. They want to get back at somebody, and the only southerners they know of who call people "nigger" are the Texans who hang out on Fifth Amendment.

"Come on, let's do something to these guys," Julio says.

And John agrees that it's time to make their move. Abandoning their handles of Corrupt and Outlaw, they don their alter-ego personas Broken Leg and Flaming Carrot—for it was they who had needled LOD members—and creep stealthily back into the Texans' seat of power. They storm Fifth Amendment.

There happened to be a particularly juicy phile posted on Fifth Amendment. It was a technical treatise on how to crack a Rolm-made PBX phone system.

Rolm PBXs, as any hacker in 1990 could have told you, are valuable properties. They are favored by large corporations and government agencies, which need to route thousands of calls a day to their employees. The phile on Fifth Amendment said that this particular type of PBX had a back-door entrance to enable repairs by remote maintenance support staff. The phile went on to explain how to break into the system using the back door, and how, once inside, to monitor phone lines. Armed with such knowledge, virtually anyone anywhere could crack a Rolm PBX.

The patrons of the Fifth Amendment board didn't want such information widely disseminated for one simple reason. If lamers got hold of it, they'd go into a PBX and screw something up. Rolm would get complaints from customers, and the whole system would be overhauled to close up the holes. So, while the phile was available for perusal, you weren't supposed to copy it. Don't share it with anybody who didn't have access to Fifth Amendment.

What better way to tweak the Texans than to spread their precious secrets?

John, acting as Broken Leg, snatched the phile, surreptitiously copying it right off its super-secure resting place on the BBS. It was a fine deception.

Less than a week later, the Rolm Revelation was posted on bulletin boards all over the country. The Texans were incensed. Such highly specific, technical information could only have come from one source: a leak from one of Fifth Amendment's users. Who among the chosen could have been so disloyal?

Scott started to monitor all the users, and within days, he noticed a pattern. Whenever John logged in, he copied philes, then logged back out. Scott and Chris immediately suspected that John was operating on the instructions of someone else, a mastermind who lurked in the background and directed John to do his bidding. Scott and Chris had the same thought—Mark Abene.

This suspicion they could not abide. Scott confronted John, electronic style. The next time John logged in to Fifth Amendment using the Corrupt handle, he saw this message flash across his screen:

```
APPARENTLY CORRUPT HAS BEEN
DISTRIBUTING INFORMATION POSTED ON THIS
BOARD TO HIS LITTLE BUDDIES. WE HAVE
EVIDENCE.
```

The battle lines were drawn now. No matter that Mark Abene had nothing to do with the Fifth Amendment escapade. Scott and Chris didn't care any more about the specifics than John did the day he heard someone calling him "nigger" over a conference bridge.

It was New York City against Texas.

Chris decided it was time to improve the image of LOD. In the aftermath of the raids, a lot of its core hackers had just plain disappeared. Maybe they'd gotten jobs. Maybe they'd gone to college. Or maybe they just got scared. In any case, Chris wanted to attract new talent, and use that talent toward "positive knowledge."

It was all part of his plan to change his life. The raid at his house had really spooked him, and his grades were going to hell this semester because he was so bummed; so it was only natural that Erik Bloodaxe decided to take action. There was no future in the underground, any fool could see it was a bunch of kids living dangerously. And Chris was getting a little old for that break-into-the-phone-system-and-control-the-world fantasy. He only saw one choice for him.

He had to change the way he viewed computers, because he couldn't *give up* computers. He spent twelve hours a day at the keyboard, what was he going to do without it? He had no other hobbies, wasn't interested in much else, didn't even believe in politics or the electoral college.

His plan was simply to stop fighting the establishment. To join it instead and beat on your enemies at the same time. Ever since he'd read a book by Donn Parker, *Fighting Computer Crime,* he'd been nursing the idea. You could get paid for this stuff? Chris thought. It was a great book. He'd bought it in hardcover.

In his plan, Chris saw himself evolving. No longer was he a threat to corporate computer security. No longer was the LOD a dark gang of anarchic kids trying to subvert phone company security efforts. No, in Chris's plan, he and LOD emerged from this bad period to become the new force for good, the hackers in the white hats who would police cyberspace for the rest of us.

He liked that vision.

Chris could see himself tracking down errant hackers—like the bad boys of MOD.

He could see himself making the judgment about whether specific hacking activities are worthy or dangerous.

He could even see himself turning people in. Yeah, he definitely could see that, because the whole scene has been getting . . . out of control.

There was a career in this. Chris began to tell people he'd like to get into computer security. He'd be a hired gun. The idea appealed immediately to Scott Chasin, who'd wanted to be a computer consultant since the age of twelve. It probably would have appealed to just about any teenage hacker in America, because in each dirty-sock-strewn bedroom sits a boy who nurses a dream of being "discovered" and offered a job doing computer security.

The dream goes like this: ENTER COMPANY NAME HERE will be so impressed with my abilities to crack its system that the CEO will call me up and offer to put me on the payroll. I'll show them how to plug the holes and to keep other people just like me from penetrating the system.

A lot of boys were waiting for the phone to ring (or would be, if they were *off* the phone). Erik Bloodaxe didn't wait. He put phase one of his plan into action immediately: reform LOD's image with new blood so the gang could earn the kind of squeaky-clean reputation that the Boy Scouts used to enjoy.

The transformation of LOD was cause for talk. Every hacker on every bulletin board seemed to know about it, and of course Mark Abene got word, too. He couldn't get back into any LOD controlled by Chris, that was for sure, but at this point he didn't even want to. He had new friends, for one thing. In the aftermath of the raid on his house, Mark was a little skittish. That didn't stop him from hacking. Mark got a laptop computer in a trade, and started hacking from phone booths, even allowing a couple of writers from *Esquire* to tag along for the sessions. But Mark was more than a little wary about being identified with an organized hacking group. To him, MOD had always been a joke. To try to resurrect the carcass of an actual, well-known gang like LOD, well, that seemed like waving something in the face of the government. You were asking for trouble.

But Mark would not be Mark if he kept his opinions to himself. Mark counseled other hackers not to join LOD: "It's boneheaded, you'll get into trouble." LOD doesn't know squat. They have nothing but moronic, out-of-date philes.

When Chris discovered the propaganda campaign, he was enraged, because he saw Mark's comments as part of a bigger pattern of harassment. You see, someone had been messing around with Chris on the phone. Chris suspected MOD. Who else could it be?

Well, in fact, it was the work of John Lee.

John never runs from a fight. One thing he likes is to see his enemy cry. He's not sure why, but he really enjoys that.

Chris is John's enemy, will be forever. But at this point he doesn't even know him by the name Chris. He just knows Erik Bloodaxe. John had decided to make his enemy's life miserable. First John had to learn Erik Bloodaxe's real name. You couldn't exactly call directory assistance in Austin and ask for a listing for a resident named Bloodaxe—that's "axe" with an "e." But Chris was so notorious in the underground that it didn't take John long to get the information he needed.

John bypassed directory assistance altogether. He instead called a Southwestern Bell computer, from there logged in to a switch, and simply looked up Chris's phone number for the three-bedroom suburban-type house he was renting in north Austin.

Then the calls started.

Sometimes John uses his ghetto accent to harass Chris. Other times, John and a friend call Chris and whisper into the receiver, "Mr. Elite, Mr. Elite," in a very mocking tone. Chris suspects that it's John and Julio, but he can't prove it, and the calls drive him crazy. Part of the problem is that Chris can't make out the words for sure and sometimes the ominous whisperers sound like they're saying "Mystery Elite, Mystery Elite." Which makes no sense to Chris.

The phone calls are constant. It doesn't help to hang up. The receiver is barely down before the phone rings again. And again. And again. You have to take it off the hook, and leave it off the hook for hours. Sometimes, when they prank Chris, the callers say, "Here, talk to your friend," and then before Chris can hang up, he hears a click, and then Scott is on the line, too, three-wayed into the call against his will, and he's saying, "Hello? Hello? Who is this?"

In Chris's mind, this type of harassment definitely fell into the category of behavior that is unacceptable. If he were already working in computer security, he could put a stop to it. He and Scott talked about the situation a lot; they even came up with a name for the company they wanted to create.

Comsec Data Security was the full, stuffy name, but neither of them ever thought of it in that formal way. No, for Chris and Scott, the venture would always simply be Comsec.

In December, a couple of momentous things happened. First, Scott and Chris met face-to-face for the first time. Whether they looked like what the other expected neither ever says. Chris is lanky and long-haired, looks like a hippie. And Scott is shorter and wiry, hair cut close like an army cadet.

The two met at Ho-Ho Con, an annual hackers' conference that kids travel from all over the country to attend. They're the new iteration of Shriners, and every year they converge in Texas for the purpose of holing up in the cheapest hotel in Houston that will take them, right before Christmas (thus, Ho-Ho). They stay up all night, hacking and talking Acronym to one another. They sleep all day, and wake toward dusk to forage for Tastee cakes, colas, and similar sustenance.

Ho-Ho Con is invitation only, but it still gets a little rowdy, and midway through the so-called conference this year, the group got evicted from La Quinta Travelodge and checked in to Howard Johnson's. Thus, Ho-Ho-Ho-Jo.

Chris and Scott were telling everybody they met about Comsec, about their vision for its future, and right away they hooked up with another hacker named Kenyon Shulman. He lives in Houston, and he's a lot richer than the other hackers. In fact, his mom is a genuine socialite, and before his arrival at the conference, the rumor was that Kenyon drives a black BMW.

When he pulled up, he was in a Mercedes. Just a Mercedes.

Kenyon was pretty excited about the whole Comsec scheme, in fact he had the deep pockets that might actually transform the idea into reality. Kenyon has an old friend in Atlanta who was about to graduate from Emory University. Kenyon's friend was headed for a job trading securities, and the two had been toying with the idea of a little startup venture.

Kenyon got on the phone at Ho-Jo's, called down to Atlanta, and before anybody knew it, Comsec was a reality.

The four of them—Chris, Scott, Kenyon, and Rob in Atlanta—figured that they'd need about a hundred thousand dollars for startup costs, but that was no problem, because Kenyon was willing to put down eighty thousand as a basis to get a loan. They figured they'd need an office to set up business, but that was no problem because Kenyon's mom would give them free footage in some real estate she owns, a one-story office building behind a strip mall in Houston.

Can you imagine how exciting this all sounded to a bunch of students? It was really great to have so many options in life, to know the right people, to have access to money when you needed it. It was the end of the 1980s, and the news hadn't sunk in yet that the decade's great leverage and expansion possibilities were threatened. No, it was still the land of opportunity out there, and all it took to capitalize on it was a few bright ideas hatched in a Howard Johnson's. Oh yeah, and a lot of cash.

But that was no problem if you made your new friend Kenyon an equal partner.

After Ho-Ho Con ended, Chris had to go back to Austin to finish up the spring semester. But Scott wasn't a student, he was working in Houston at what he called a "poor-slob job." He entered data all day into the computers of the Academy Corporation, which owns a chain of sporting-good stores throughout the state. Scott and Kenyon were working all night, every night, to lay the foundation for Comsec before Rob and Chris moved to town in the spring.

Scott went to the library to look up all the computer security firms in the United States and counted them. Not too many. He researched what each company specialized in, analyzing the industry to see how best to position Comsec. Research, research, every night. He even researched how to buy office supplies, where to get the best deal.

Of course, there is one major distraction: the MOD-LOD war, which reaches a crescendo in the winter of 1990.

One day, Chris gets hold of a copy of "The History of MOD." The spurned Alfredo gives it to him, proving again there is a high price to pay for adolescent cruelty.

Chris feels he has been teased and provoked enough; the MOD boys have logged into the Southwestern Bell switch that controls his phone service and switched his long-distance carrier from Sprint to AT&T. Chris didn't know that had happened until he tried to dial long-distance. He doesn't hear the familiar click. So then, of course, he had to call up the phone company. Try explaining the situation to a clerk in the business office and you'll know why he's so annoyed.

Chris figures that John was the one who switched his long-distance carrier on him. He also believes, incorrectly, that John is the author of "The History of MOD." So Chris gets hold of the Boswellian tale and decides to pull a little mischief.

Chris has an old computer program that will translate any file into a new "language." In this case, when he feeds "The History of MOD" to the program, out pops a "jived" version of the document. The program simply searches for certain words or word forms, and replaces them with others.

In goes the original language that Eli wrote: "In the early part of 1987, there were numerous amounts of busts in the U.S. and in New York in particular."

Out comes, "In de early part uh 1987, dere wuz numerous amonts uh busts in de U.S. and in New Yo'k in particular."

In goes Eli's description of his own activities in 1989: "It came about when Acid Phreak, then using another handle, had been running a semi-private bbs off his Commodore piece of shit and ten generic Commie drives."

Out comes the jived version: "It came about when Acid Phreak, den usin' anoda' handle, had been runnin' a semi-private fuckin' bbs off his Commodo'e piece uh shit and 10 generic Commie drives."

The jive program also adds racially offensive phrases to the text.

Every couple of lines, for instance, the phrase "slap mah 'fro" or "sheeeiit" appears in the text as a kind of hateful refrain: "Hims handle wuz De Win', and he ran some Unix systum fum his crib in Pennsylvania. Sheeeiit. Sco'pion wuz always some Unix guru while Acid had only jest across it in college two years back."

Using the jive program is the electronic equivalent of appearing in blackface.

Chris does his work in the gracious surroundings of Kenyon's townhouse. An English tapestry depicts a hunt scene on one wall of the office, bookshelves line another. On the shelves, looking down at Chris's work, are the unjived versions of books by Hume, Freud, Plato. It takes Chris a couple of hours to get the jive program to work right, to convince it to insert spaces in the right spots between words, and to add punctuation where called for. He works intently, enveloped by a red leather chair that's roomy enough to sleep in: "Some nigga' name Co'rupt, havin' been real active befo'e, duzn't gots' some so'kin' computa' anymo'e and so . . . sheeit, duh."

Chris doesn't consider himself a racist. He has black friends at work, he says. If you ask him why he jived "The History of MOD," he says it just seemed funny. Hilarious, he says. If you're out to get someone, you're going to do anything you can to make him mad, Chris says. Anything. He didn't have a translation program to turn the MOD boys' prose into, say, a *Lithuanian* accent or something, he only had a jive program. So what was he supposed to do?

If you lived in Texas, you'd understand, Chris says. "Down here, we all have boots and hats. We all ride on the range."

Long before the jive version gets officially published in the April 1991 issue of *Phrack,* John sees a copy of Chris's handiwork.

Of course, John isn't sitting in a red leather chair in a room adorned by tapestries when he sees it. He's sitting in front of a computer system that looks like it was cobbled together from junkyard parts. He has a big old TV console for a monitor, a messed-up keyboard, and his old Commie 64, bandaged with electrical tape. His computer is a street box, a guerrilla machine.

It's about seven in the evening, dark enough already to warn you away from ideas that spring might come early to New York City, and John's sitting in the same chair he always uses at the computer in his bedroom. From time to time, he hears sirens scream down faraway streets. He can also hear, through the open windows, the rapacious demands of rap music. The lyrics and the insistent beat drift up from the sidewalk, as one boombox after another passes under the window of his row house. Inside, the only things on John's walls are a couple of posters advertising NYNEX yellow pages. Their messages are funny, and that's why John has them on the wall. One shows Barbie and Ken looking under the entry for "Plastic surgeons." Another shows the entry "Chicken ready to serve" under a picture of a hen with a tennis racket.

John has a third handle, Netw1z (Corrupt and Broken Leg somehow just don't say it all for him). He sees that a file has been sent to his Netw1z account. And there it is on the screen, the jived "History of MOD."

"De legacy uh de underground 'clandestine' netwo'k continues and so's duz de war (and ridiculing) against all de self-proclaimed, so-called 'elite.'"

John can't believe it at first, it's too outlandish. He reads through it, slowly, amazed. *Slap mah 'fro.*

John finishes reading, then sits for a minute, staring at the screen, staring away from the screen—just kind of staring. And he thinks, This guy really doesn't like me. This is aimed right at me, and only me.

ELEVEN

One day Tom Kaiser's phone rings, and on the other end of the line is an official from Tymnet.

Tymnet, you will recall, is the sprawling, privately owned data network that computers use to send information to one another around the world. You dial a local phone number. It's called a dialup, because you're literally dialing up the Tymnet computer. Once you're in, you can connect to any computer worldwide that's hooked up to the system.

Today, the Tymnet official tells Kaiser that Tymnet has a problem. The huge network has somehow been penetrated by an unknown hacker. The hacker is roaming through computers that belong to a rather influential customer. The customer's computers are all hooked up to one another, forming what is known as a subnetwork within the Tymnet structure. The official says Tymnet has only been able to trace back to the general area where the calls are originating: New York City.

Oh, and one other thing. The Tymnet official would rather not identify the customer. It's a sensitive situation. You understand.

Kaiser understands sensitive situations. The lawman says he thinks he can help.

The trick here will be to figure out where the intruder lives. Tymnet traced the call as far back as the New York City dialup that the hacker used to enter the Tymnet system. Of course, Tymnet has hundreds of dialups in major cities worldwide, and more than one in New York City. So even if the hacker phoned a specific New York City dialup one day, he could easily call another one the next time. There's no way that Kaiser can instruct dozens of phone switches around the country to monitor, or trap, all the calls coming in to every single one of Tymnet's phone numbers.

Kaiser, however, can try to do a live trace while the hacker is on the phone. That's dicier, because if the hacker hangs up before the trace is completed, the intruder is gone. There's no way to say where he called from.

Kaiser has no clue about the culprit's identity. The last people he suspects are members of the MOD squad. After all, a year after the raids, cases are still pending against Mark, Eli, and Paul. It's winter again in New York City, a slushy yucky affair that's been slopping over the tops of Kaiser's shoes every morning as he tromps up the steps from the subway into his office building. It's been a little frustrating for the lawman, waiting for the federal government to resolve the hacker cases. In fact, the assistant U.S. Attorney in the Eastern District who handles the cases has been running into some roadblocks. It seems that his bosses don't think that prosecuting a handful of teenagers is a good thing to do with the resources of the office. It seems that the bosses don't really grasp the significance of the case. "What's a switch?" they ask.

The boys' cases had been bounced out of federal court, in fact, and handed off to the Queens District Attorney. There's been a little delay as that office got up to speed. Kaiser even made a few trips out to Queens to explain the technical aspects of the case. Finally, charges were filed against Mark in state court, charging him with a misdemeanor.

So Kaiser certainly hasn't forgotten about the MOD investigation. Far from it. One worry he harbors is that hackers might harm the New York Telephone system in retaliation for the MOD prosecution. There's been nothing Kaiser can do to relieve his worry, though. He has no proof. No evidence. No tips. Nor does he have any reassurances that the hackers won't retaliate. While the phone company has a legal right to monitor its own lines, Kaiser can't abuse the privilege by randomly spying on people. Before he asks his bosses for permission to put up a DNR, he has to have convincing evidence that such a step is necessary. He can't say he wants to monitor a customer's line simply because he *worries*. He needs some concrete reason to do it.

On the phone with Tymnet, Kaiser goes over the procedure for a "live" trace: As soon as the intruder is spotted in your network, call me. I'll start to track him immediately. It's essential to move fast. If the connection is broken before we complete the trace, we'll never know who the intruder is.

Got it.

By the time Kaiser learned who Tymnet's secret customer was, winter had given way to spring. It is May now. It's warm, it's sunny, you can even hear birds tweeting in Manhattan, for God's sake.

Kaiser had been expecting to hear from Tymnet again. But this time, the phone call comes directly from the Tymnet customer, a client who well knows that minutes are precious during a trace. The customer is Southwestern Bell.

Southwestern Bell is a sibling of New York Telephone, another of the regional phone companies born from the AT&T divestiture. But Southwestern Bell can't execute a trace over New York Telephone's lines, having no access to phone lines in the Northeast. Southwestern Bell's kingdom stretches across Arkansas, Kansas, Missouri, Oklahoma, and Texas, where a web of computers connects about twelve million customers. Those computers perform the company's switching operations, route calls, bill customers, and process administrative work.

Since its computer system spans so much of the country, Southwestern Bell decided to set up a subnetwork on Tymnet. That way any Southwestern Bell employee could call into the system for the price of a local call, using a Tymnet dialup. The Tymnet pipeline is a great convenience to Southwestern Bell's employees.

It was also a great convenience to hackers. That pipeline enabled a hacker sitting anywhere in the United States to call a local Tymnet phone number, get into Tymnet's system, and then burrow right into Southwestern Bell's computers. Southwestern Bell became aware of the problem one morning when the administrator for one of the company's central depots, the C-SCANS system, came in to work. The administrator noticed from a log that it appeared he himself had been signed on and using the computer during the previous night. He had not. Now, C-SCANS is a sensitive and complex system. The Client Systems Computer Access Network Standards is a central operation that distributes information and administers patches throughout Southwestern Bell's region. When the Bell company wants to upgrade software across its network, the new program is funneled through C-SCANS into each switch. From C-SCANS, you can hop right into any of those switches and start looking around. C-SCANS also stores internal electronic mail sent among Southwestern Bell workers. Security memos are stored there, too.

As soon as the system administrator realized that some un-authorized user had penetrated the system, he had to bring down the computer to try to find out how the intrusions occurred. Bringing down the computer costs Southwestern Bell money, both for paying employees to fix the problem and in lost computer processing time. But the hacker didn't seem to be limited to any one spot in the system. Perhaps because he'd learned so much about the other switches he explored through C-SCANS, he was popping up all over Southwestern Bell's computers. The people at Southwestern Bell sure would appreciate Kaiser's help.

And today, on May 31, 1991, at this very moment of 4:31 in the afternoon, a hacker is logging into Southwestern Bell's mighty Netcon VAX system in St. Louis, which controls switching networks in four states.

The hacker has a valid user ID: "Carolw." The "Carolw" account belongs to a communications technician with access to sensitive documents about network security within the Netcon VAX system. The technician doesn't know who could have gotten hold of her ID.

"It's urgent, he's on the line right now," the Southwestern Bell official tells Kaiser.

Southwestern Bell reports that the call came in through Tymnet. And so Kaiser keeps Southwestern Bell on one line, then immediately calls Tymnet officials on another. Let's say the hacker got into Tymnet by using the dialup at 555-4700.

"That's a number in Lower Manhattan," Kaiser says, recognizing the prefix. The drill is to trace the call backward from the dialup to its origin.

Minutes are ticking by as Kaiser puts Tymnet on hold and dials the New York Telephone switching control center located downtown on West Street.

"This is Tom Kaiser in security. Where is the call coming from?"

Down in the West Street office, nobody hesitates. The technicians know that when security calls, it's not for a run-of-the-mill request. No, West Street *might* put Kaiser on hold for a fraction of a second and simultaneously call him back on another line to verify he's really who he says he is. But by the time Kaiser answers that call (he has a lot of phone lines in his office), West Street will have the answer he needs.

"The call's coming in on an AT&T trunk line," says West Street.

If you make a long-distance call using AT&T, your call travels to its destination along a line that's designated to carry AT&T customers' calls. That's a trunk line.

"AT&T?" Kaiser repeats, and he's already dialing again. This is bad, since it means the hacker may not be calling from New York City after all. He could be anywhere, just using the New York City dialup.

Meanwhile, back at Southwestern Bell, security personnel audit the intruder's every move as it occurs, hoping the hacker won't get bored and hang up. It's maddening, not to mention terrifying, since the phones in four states are controlled by this VAX computer. They watch in shock as the hacker, disguised as Carolw, coolly calls up some internal security alerts sent by Bellcore. Bellcore, or Bell Communications Research, is the research and development arm of the seven regional Bell companies. Back in the 1980s when AT&T's monolithic empire was broken up—divided into a long-distance company that AT&T could still control and seven autonomous regional siblings—one thing AT&T was allowed to keep was the prestigious Bell Labs. The telecommunications laboratory has an illustrious history. Researchers at Bell Labs developed everything from the laser to talking motion pictures.

The Baby Bells got together and funded their own think tank, their own Disneyland of future ideas: Bellcore. Much of what Bellcore does is develop new standards for improving the phone networks. Bellcore researchers also do a lot of on-the-edge, blue-sky stuff, trying to figure out the future direction of the communications revolution.

One of Bellcore's most important functions is security. Bellcore periodically sends memos to all the regional telephone companies to announce a breach of security at one or another of the local phone companies. The memos describe intrusions, and in some cases, the intruders themselves. To a hacker this is extremely valuable information, because the memos also explain security techniques that exist to fix the loopholes. The hackers stay a step ahead.

This particular hacker is browsing in those memos now, reading files about other hackers. He's hacking Southwestern Bell's anti-hacker archives.

It's almost five o'clock in New York, and nobody knows how much longer the call will last. Kaiser has AT&T security on the phone now, and the long-distance carrier's technician quickly traces the call on trunk line to a toll-free 800 number in Pennsylvania. This can mean only one thing. The hacker is purposely hiding his tracks.

Here's how. The hacker had called a toll-free number assigned to some unsuspecting corporation in Pennsylvania, the same toll-free number the corporation's employees use to check in. The computer has a PBX, a private branch exchange to manage internal phone calls, only like most of the novices in telecommunications, the company doesn't know how to secure the system. Who would suspect that someone would want to breach it? Ah, but lots of hackers have found these vulnerabilities and were using them like free long-distance calling cards. This hacker merely pressed 9 to get a dial tone that would allow him to call back out, right from the corporation's phone system, to Tymnet. It's simple. And it's smart, because it's certainly slowing down Kaiser.

Kaiser checks his watch as he dials the officials at the Bell Atlantic switch that controls the (800) number.

"Where's the call coming from into that (800) line?" he asks.

"The call is coming from New York."

New York. Back in Kaiser's jurisdiction again.

Even as Kaiser works the phones faster than Lily Tomlin did on "Laugh-In," officials at Southwestern Bell are getting antsy, sending a collective mental plea to the hacker. Please don't hang up. Not yet.

"Where in New York?" Kaiser thinks he's yelling into the phone, but really, his tone is quite calm. He's only shrieking *inside*.

"Brooklyn."

Kaiser calls the New York Telephone switch in Brooklyn.

Now Southwestern Bell officials are talking in his other ear, saying the intruder has finished reading the memos. He's copying the files onto his own computer to read later and divide like stolen money. Is the hacker getting ready to hang up?

Into the other line, Kaiser barks at a technician at the Brooklyn switch, "Kaiser from security. Where is this call originating? I need to know *now*." The New York Telephone technician says, "The call is coming from 555-1318."

Success.

It's 5:09 P.M. The trace took thirty-eight minutes, much longer than a trace normally takes. But then, a typical trace request is one that comes from the cops, who say there's a potential suicide on the line. Where's he calling from? That's an easy enough matter—just call the switch, ask one question, write down one number. Five minutes, you're finished. Kaiser's done it plenty. Today's activities, however, are of another magnitude.

Within a mere thirty-eight minutes, Kaiser has traced a call back from Southwestern Bell, through Tymnet, through a New York Telephone switch to an AT&T line, to a toll-free number in Pennsylvania, to a switch in Pennsylvania, then back to another switch in New York City, and finally, to a building in Brooklyn. The hacker threaded his call, in an attempt to escape detection, through at least six separate computers. And Kaiser smoothed out the tangles. That is a day's work.

Kaiser writes down the number, and then he looks it up on his own computer. Let's see, 555-1318 is assigned to a subscriber at 64A Kosciusko Street, in Brooklyn. Weird, it even sounds familiar. The hacker lives in the neighborhood of Bedford-Stuyvesant.

Kaiser gets back on the phone to Southwestern Bell's anxious officials and gives them the news. He hears a cheer go up on their end of the line.

Kaiser remembered where he'd seen that address and phone number before. John's number was one that Mark Abene called frequently back in 1989 and early 1990. The DNR kept track. Kaiser had seen John's name before, too, scrawled alongside the doodles in a notebook confiscated from Mark's house during the raids.

That's how Kaiser knew John Lee's handle was Corrupt. That's how Kaiser knew John Lee was in MOD.

In the days that followed, Kaiser would perform more traces for Southwestern Bell and Tymnet. Sometimes the connection got broken before he could navigate the labyrinth of phone lines. But Kaiser was successful in tracing another call, this time to Julio's address in the Bronx.

Julio's phone number was in Mark's notebook, too. So was Julio's handle, Outlaw. Julio must be in MOD, too.

So Southwestern Bell knew that it had to deal with at least two intruders. And now Kaiser knew that he was back on MOD's trail.

The Secret Service was moving on the case. Big time. But the Secret Service was also starting a new investigation, treating the Southwestern Bell intrusions—which clearly posed a threat to a significant portion of the nation's phone lines—as a case of their own. The Secret Service didn't put it together with the old MOD case from 1990. As a new case, the Southwestern Bell problem has a new prosecutor. It's the U.S. Attorney in the Southern District, which has geographic jurisdiction over Manhattan and the Bronx.

This was the first computer crime case assigned to Stephen Fishbein in the three years he's been working as an assistant U.S. Attorney. But he is far from technophobic. In fact, Fishbein has an identical twin who is a chip designer in Boston, and Fishbein is not afraid of computers. That put him a step ahead of some of the other lawyers who would find themselves involved in what would become known as the Masters of Deception conspiracy case.

But even after Fishbein started delving into the computer intrusions that John and Julio committed from their homes, it would be months before the full scope of their involvement with the earlier case against Mark, Paul, and Eli became clear.

After the successful traces, the authorities tell the switch that controls certain phone lines to Southwestern Bell's computers to trap incoming calls. Whenever someone calls the phone numbers, a computer notes the origin of the call. Every time John or Julio calls, a computer records the time and date. The monitoring is quite definitive.

DNRs go up again, this time on the phones at John's and Julio's houses. Here's what they show: On June 2, at two minutes before midnight, Julio calls Tymnet for nineteen minutes. As soon as he hangs up, he calls Mark.

The next day, around suppertime, Julio makes a one-minute call to Tymnet, then a fifteen-minute call to another Tymnet dialup. Then he calls Mark.

On July 1, in the early afternoon, Julio calls Mark twice. Two minutes later Julio calls Tymnet for nineteen minutes. Then he calls Mark.

On July 4, at four in the afternoon, John makes nine calls in a row to Tymnet. Then he calls Mark.

Over and over again, the pattern repeats, just like it did when the earlier DNRs were up in 1989. Calls to Tymnet, calls to Mark. Calls to Tymnet, calls to Mark.

You'd think Mark would know better than to engage in illegal hacking by now. After he was charged with a misdemeanor in State Court in Queens, he pleaded guilty to call-forwarding phones to a 900 number. Mark's plea did not specify the nature of the 900 number, but police told reporters that he was calling sex lines. Mark hotly denied it—what would he be doing calling sex lines?—but there was nothing he could do to quell the rumors. Mark's sentence was thirty-five hours of community service at a hospital in Queens.

The outcome of that case reinforced Mark's perception that hacking was not an offense the government took seriously. Despite being convicted, nothing really bad had happened to him. Somewhere between early 1990 and now the case had fizzled. Not only had the U.S. Attorney's Eastern District office wrangled out of prosecuting the kids but the Electronic Frontier Foundation had never followed up on its initial interest in the case, a sign the hackers interpreted as meaning they weren't in real danger from the government. The MOD boys didn't know that Kapor and Barlow backed off because Kapor's lawyer had told them he didn't see a real civil liberties case to pursue. The boys' 1989 and 1990 transgressions appeared to them to be petty trespasses.

So it was no wonder that Mark rationalized his decision to

continue hacking. That was his identity. Phiber Optik. No where else could he get this kind of intellectual stimulation.

Why don't you just go to school, Mark? Why don't you just complete your high school degree and talk your way into a nice computer science department somewhere?

School is not for Mark, he says. They don't teach the kinds of things he wants to know. Mark only cares about specific, arcane computer networks, the crazy puzzling over how they're put together and why they work a certain way. Phone company networks. Tymnet's system. He can lecture for hours about those systems. Mark's a specialist who isn't interested in the more general knowledge that computer science departments offer about how to program a VAX or a Unix computer. No, Mark is as niche-oriented as a medieval scholar.

So why doesn't he get a job with a phone company? Well, there's this misdemeanor conviction, for one thing. But for Mark, another bigger reason is that he believes that if he went to work for a phone company, his understanding would be limited to the architecture of a single, local system. That's not for Mark. He wants to know how *everything* works.

And he's willing to teach anybody else.

Which is why he is leading Julio and John through the intricacies of navigating Tymnet to get to Southwestern Bell. They are eager students.

Why are John and Julio so interested in this endeavor? Southwestern Bell's C-SCANS controls every switch in Texas. Southwestern Bell controls the phone service of Chris Goggans, of Scott Chasin, and now, of the brand-new Comsec offices.

TWELVE

Kenyon's mom hired some people to fix it up, so the new office was really cool. Very professional.

Comsec opened its doors in May of 1991, and the business partners quickly made themselves at home in the airy headquarters. Comsec has a huge vaulted ceiling with skylights and faux gaslights in the two corridors. Some days Comsec's founders skateboard down the long empty halls of the vast space or roll around in chairs. Chris was living in the back of the building, in an apartment with a big white-bellied alley cat named Spud.

But there was a problem.

Comsec had zero clients.

The officers of Comsec hold weekly staff meetings, which they all attend. They decide to distribute press releases advertising the availability of their security services. But to whom? As ex-hackers, they compile a list of likely clients. They scan the philes on underground bulletin boards to find the names of businesses whose computers have been infiltrated, then call the companies to offer their services.

The press releases must have done the job, because in June, less than a month after Comsec officially opened its doors, both *Time* and *Newsweek* ran stories about the hackers-turned-anti-hackers.

The very next day the office phone started to ring. And ring. You couldn't buy advertising better than *Time* and *Newsweek*. Comsec has clients! One client, a consultant representing the telecommunications industry, ordered up some research on recent regional Bell company crashes. The client paid five thousand dollars up front.

Of course, the publicity in *Time* and *Newsweek* had another effect.

Up north, the MOD boys are reading the stories.

The pay phones are jammed.

So many hackers mill around the kiosk in the Citicorp atrium on this afternoon that you can't even see the phones.

It's the first Friday of the month, and most of the MOD boys are at the *2600* meeting. Even Paul's here, home from college and hanging out. Somebody has arrived at the *2600* meeting carrying the issue of *Newsweek* with the article on Comsec, and the magazine gets passed around until it's downright grubby. The articles foment this sort of bad feeling, an unsettled not-quite-rage, not-quite-amusement, that they feel in their guts. Then some hacker says, "Who do those guys think they are?"

Spoofing hackers is one thing. While it may hurt some feelings, no one gets really jammed up by it. But what kind of a hacker turns on other hackers? It's a question that no one can answer.

Suddenly, the hackers rush the pay phones in Citicorp, and in each little carrel, they launch a mass attack. Everyone dials the same Texas phone number: Comsec's 800 line.

As a result, down in Houston, the phone bank lights up. Every line rings at once, and the Comsec crew doesn't know which phone to answer first. Of course, it doesn't really matter which one they answer, because all the taunting callers say the exact same thing, in the exact same maddening sing-song tone: "Hello? Is this Comsec? I have a victim here who wants to hire you. . . ."

All afternoon, the lines stay busy at Comsec, so no legitimate client could possibly hope to get through. As soon as the Comsec crew hangs up on one caller, the phone lights up again. The kids at Citicorp have found an effective way to communicate their displeasure.

The electronic heckling doesn't abate until the Manhattan "meeting" breaks up at about 8:30 P.M. Eastern Standard Time. The kids wander off in little clumps, headed downtown to grab a cheap supper and to browse around in record stores. They have their monthly routine.

First, they go to Around the Clock, a moody, dark East Village restaurant where rap music plays on the juke, a big color TV blares on the wall, and athletic waiters wear knit, olive-drab rave caps. The food is just this side of hippie: wholewheat pita concoctions, three kinds of organic pancakes, "healthy" chicken soup. Mark usually orders two bowls. It settles his stomach.

After eating, the boys wander across Third Avenue to Tower Books, where they flip through the endless shelves of hardcovers. A hacker sees a phone receiver hanging on the wall and starts fiddling. It's the store's internal intercom system, and it can't make calls outside the building. But with a device he happens to carry, the hacker coaxes a dial tone out of the receiver and immediately places a call to a friend in Pennsylvania. The device he uses is a tone dialer, which emits a noise that simulates the sound a pay phone would recognize as coins dropping into a slot. It works like a bird call for computers. It was a pretty good hack, you'd have to admit, until all of a sudden a security guard comes over. He starts hassling them, wondering why some scruffy kid is fooling around with the store's phone. At that point, things take on the flavor of a segment from the Keystone Kops. Hackers scramble for the store's exit, more security guards appear from nowhere. Then another hacker gets this bright idea to spray a Mace-like substance into the air as if it's air freshener. The hackers are pumped now, and they rush out into the New York City night. Nobody gets caught. They have a good laugh about it.

.annoy

That's how you type it: period-no space-lowercase annoy.
Dot-annoy.
That's how all the MOD boys say it. *Dot*-annoy. And they should know, since they created the simple string of Unix commands that comprise ".annoy."
In the MOD vernacular, dot-annoy is both a noun and a verb. Here's an example of how you dot-annoy someone else.
First, log on to Allen's Unix computer. Then leapfrog over to the Apple computer that's hooked up to it.
Next, pick a lamer who deserves a taste of dot-annoy.
Find his name in the database.

Find his name in the database.

Call up the lamer's entry and execute the dot-annoy program. It's easy. This will instruct the Apple computer to follow the simple software program that Allen wrote. Let's say the lamer's phone number is (212) 555-1234. You simply type in the phone number, and the program does the rest:

```
while :
do
cu 12125551234
done
```

Looks pretty innocuous. But the purpose is anything but. This little string of commands instructs Allen's computer to use its modem to call the lamer's phone.

Now, you can expect the lamer to answer the phone, then hang it up because no one is on the other end.

This is where dot-annoy gets ingenious.

The MOD users add a simple command to the dot-annoy program to make it repeat, endlessly. Over and over the computer will redial. All you have to type, at the top of the dot-annoy commands, is this:

```
nohup .annoy&
```

That command tells the computer the program does not end at a hang-up. Accept no hang-up. Keep dialing until another modem answers. So, as soon as the lamer's receiver returns to the cradle, the Apple will call him again, looking for a modem. It has to.

The lamer will answer again, then hang up.

The Apple will call again.

Hang up again.

Call.

Hang up.

Call.

Hang up.

Call.

Call.

Call.

Call.

Execute this plan at the beginning of a long holiday weekend, say Fourth of July, because then the lamer can't seek assistance from the business office *until Monday*. You won't have to do another thing, not after that initial instant when you call up the lamer's file to execute the dot-annoy command. (That ampersand at the end of the "nohup" command is Unixspeak for allowing the program to run in background—in other words, the MOD computer can be used for other things while the dot-annoy program runs.) You've automated the job of harassing your quarry.

It's going to be a long weekend for the lamer. He will go crazy. The lamer will want to cry. The lamer will have a lot of explaining to do to his parents. Even if he's smart enough to answer the phone with a modem, the torture continues. Because as soon as the modem-to-modem connection is broken, that is, as soon as the lamer hangs up, the hell starts again.

And best of all, the lamer will not be able to resist the trap. Time and again, he will pick up that ringing phone, because he will *just want to stop the noise*.

Too bad he can't.

The MOD boys could get really cute and add refinements to the torture. One twist is called "Mr. Ed." Invoke Mr. Ed, and Allen's computer would call the target and speak directly to him. Allen's computer, which has audio capability, would blare out, "Hel-l-l-l-o, Wil-l-l-l-bur!" over and over.

The MOD boys rely on the dot-annoy reportoire when dealing with the Texans, of course. But they also attack bystanders. Any hacker who has crossed them, any hacker who has the misfortune to be labeled lame, is prey. The first victim was an old Boy Scout acquaintance of Allen's, back from the days when Allen was striving for Eagle Scout. Now the database has dozens of names.

The database is the key to dot-annoy.

You can't dot-annoy someone unless you know a few salient facts: hacker handle, which leads to real name, which leads to real phone number, which leads to local phone switch. Hence, the database. The list gets a little out of control, though. The MOD boys are a little obsessive about it, investigating the people on the list and then adding to lamers' records information about street addresses, *previous* addresses, cable-and-pair number, known relatives, best friends, even anecdotal material about lamers' personal *habits*. It pays to know.

A lot of people say the MOD boys have become bullies. The boys aren't listening to that kind of sour-grapes talk, though.

The MOD database resides on Allen's Apple, accessible through Allen's Unix, which has become the MOD boys' electronic clubhouse.

Allen nas renamed his bulletin board MODNET. While every hacker in America knows about the existence of MODNET, very few have access to it. All the MOD boys have accounts, and they keep their most top-secret privileged information stored on MODNET. Everybody knows about the database. Rumors get around.

The existence of the dot-annoy database is what really freaks people out, though. What if Allen got arrested and that information fell into Secret Service hands? That file contained explicit biographical information on dozens and dozens of hackers around the country! Around the world! It was a blueprint of the whole underground. The mere possession of such damning information seemed like blackmail. It made other hackers frantic.

Any hacker would kill to get into MODNET. Good luck.

Besides the database, what was actually archived *on* MODNET was a rather eclectic compilation of data. There were loads of philes about technical aspects of hacking phone computers. There was "The History of MOD," in its entirety. There was also a "History of the Knights of Shadow," a Rosetta stone of sorts about an early band of hackers that spawned Lex Luthor and his LOD. There's a wire-service story about the disposition of the legal case against one Peter Salzman, a.k.a. Pumpkin Pete.

The MOD boys also did a fair amount of star-gazing. They kept entries on certain celebrities—those credit histories that John would look up on a whim. Geraldo's financial profile was there. So was David Duke's. So was John Gotti's and Julia Roberts's, and Winona Ryder's. *Mad Magazine* founder William Gaines was also in the pantheon of celebs because John had a vague plan to call him and ask for a job. So was Christina Applegate, a TV actress on whom Chris Goggans was said to dote. The MOD boys had been letting it be known that they'd called up Ms. Applegate pretending to be Chris.

"Maybe you saw my photo in *Newsweek,* babe," the faux Chris supposedly told her. The real Chris was just sick about it.

The real Chris, in fact, knew all about the MODNET database, all about what those New York boys were up to.

It seemed that one of Allen's pals, a Houston hacker who we'll call The Dentist, was an authorized MODNET user. The Dentist also fed information to the Texans. The Dentist was a double agent.

The Dentist told the guys at Comsec all about his account on MODNET. The Dentist was anxious to please.

"Give me the account," Chris said.

Well . . .

"Let us take a look at it," Chris said.

Well . . .

"Come on, we won't *do* anything," Chris said.

The Dentist, who'd been hanging out a lot in Comsec's offices lately, shares.

Pizza boxes and Coke cans litter the office. Full-bellied and full of steam, the Comsec boys sign onto their IBM 386 tower computer and call MODNET. Chris jokes about it: Oh, no, the lame guys from Texas break into MOD's computer! He's really pretty pleased.

The Comsec modem places the call, and on the first ring, the MODNET modem answers. Connection!

Here's what the Comsec boys see on their monitor:

```
M O D N E T
Please use lowercase when logging in.
If you are new here, log in as "new",
password "new".
Modnet !2400 baud login: dentist
dentist's Password: xxxxxx
The current time is 21:03
SECURITY Password:
UNIX System V Release 3.51m
modnet
Copyright (c) 1984, 1986, 1987, 1988,
AT&T All Rights Reserved
```

```
Last login:
No mail.

M  O  D  N  E  T  M  O  D  N  E  T  M
O                                   O
D                                   D
N     MODNET UNIX / SYSTEM V O      N
E                                   E
T                                   T
M  O  D  N  E  T  M  O  D  N  E  T  M

Your mailing address is:
modnet!dentist@uunet.uu.net
Extension 227 has 0 voice mail messages.
Extension 911 has 0 voice mail messages.
modnet$
```

That last line is the prompt. It's prompting the Comsec boys to issue a command.

Usually, the first command you type whenever you get inside a UNIX system is:

```
ls
```

That command lists all the subdirectories and files and tells whether you, the user, are authorized to read them.

Now, The Dentist does not have the highest level of access. He's not even a member of MOD, he's more of a hanger-on than anything else. Most of MODNET's supersecret delicacies are beyond his reach. He is not authorized to get into the dot-annoy database, for instance.

This minor problem is no surprise to Chris, however.

Not only did he anticipate the limited usefulness of The Dentist's account, but Chris has a plan to surmount the problem. The plan can be summarized in two little words.

Finger bug. That phrase, fraught with all the innuendo that any teenager could hope to convey, is the name of one of the better known security holes in the Unix operating system.

First, you must know what "finger" is. "Finger" is a common Unix command that tells you whether a certain user is logged in on your network. You finger, or locate, the user. For instance, if you want to find out if a user named k00ldewd is logged into the system, you type:

```
modnet$ finger k00ldewd
```

In response, you might see k00ldewd's name, his real name, the date and time he last logged in—is he logged on now?—and whether he has any unread electronic mail.

In addition, the finger bug would show you k00ldewd's .plan file (that's *dot*-plan, of course), which typically is an autobiographical description of the user you want to locate. Think of the dot-plan file as a user's high school yearbook entry.

```
i'm a k00ldewd
i love depeche mode
i drive an IBM 386 clone
my favorite food is Cheeze Waffies.
deth to lamerz and rodents!!
```

The finger *bug* exploits a well-known vulnerability in certain Unix systems. The bug temporarily gives root access to anyone invoking the finger command. Let's say you wanted to read a file called k00ldewd.mail. Using another string of commands, you just link k00ldewd.mail to .plan. Hence:

```
ln -s k00ldewd.mail .plan
```

Chris uses the finger bug to read the electronic mail of all the MOD boys. He reads all of John's mail, and then copies it for later gloating:

```
From: The Wing
To: outlaw corrupt
Subject: username
I got a username if you don't have this
one. "mcreese" p/w "blue moon" on tymnet.
also got "tnxmdhit01" p/w "ufonetran"
(jumps into some sort of xmodem shit).
```

```
Welp, I'm outta here . . .
The Wing
of M.O.D.
```

Most of the stuff, if you must know, is pretty dull, pretty weak, pretty, well, unreadable. Like most people's e-mail. But it's the principle of the thing. Comsec has broken into MOD's hideaway. Chris efficiently makes copies of all the files so that he'll have a record of his exploits. Of course, it would be awesome to have also infiltrated the database. Can you imagine? Bet you could change your own phone number so that when those New York idiots tried to dot-annoy you, they ended up calling themselves.

But the Comsec boys couldn't get into the database because, well, they couldn't even find it. They didn't know that the database is housed on a totally separate computer, the Apple. Oh well.

Chris would like to keep the buffered trophies proprietary to Comsec. They'd be useful to show to prospective clients: These kids think they're so smart, doncha know, they even have a file called Comsuc. But look at this, looks like your computer system is one of the ones they can infiltrate. Your system is one of many in a file called MODOWND. Get it? MOD-owned.

However, Kenyon has to go and slip the copies to Alfredo. Couldn't resist.

Alfredo "publishes" the scandalous information in his electronic newsletter, "The NASTY Journal." It publishes sporadically. Twice, in fact. In its final report, "NASTY Journal No. 2," publisher, editor-in-chief, and staff writer Alfredo De La Fe scores a beat with this exclusive scoop:

> During the past few months, NASTY has taken a small vacation. During that time, MOD has bragged about 'crushing NASTY with my thumb'. Well, it just got to be unbearable. It's time for us to show the little shits for what they really are.
>
> Well, let me start off by recapping the situation. MOD claims to be so dam untouchable. They also claim [MODNET] and their UNIX are so dam secure. Hehe what a joke. During the time NASTY has been 'crushed', we have been monitoring [MODNET]! All of the mail, files, password files, messages, and lovers' quarrels have been intercepted. YES we OWN

MOD! EVERYTHING, I mean EVERYTHING is going to be made public. Including, but not limited to, MOD's 'PRIVATE' database!

Nice going, Alf.

You might expect the MOD boys to be swooning from the embarrassment of it all. They don't swoon. In fact, they issue spin control on the incident. Finger bug, schminger bug. The whole thing was a trap, don't you see, to lure the Comsec crew. MOD knew The Dentist couldn't be trusted. They *wanted* Comsec to disseminate, far and wide, evidence of MOD's incredible prowess in accumulating information. Bad publicity is still publicity, isn't it?

One day, John Lee has an ingenious idea for pranking the Texans. Why didn't he think of it before?

He puts his plan into action during the long, hot summer of 1991. It keeps his mind off the lack of air conditioning in the brownstone apartment on Kosciusko Street. The mechanics of the spying are fairly simple. John logs in to the Southwestern Bell switch that controls Comsec's phone service in Houston.

Then John types commands to ask the switch if any of Comsec's phone lines are off the hook. If they are, then John would know that a conversation was under way right now.

A phone line is in fact off the hook. So he issues another command, just like an operator would, to seize control of the line that carries the call. That easily, he splices himself into the ongoing conversation.

There's a quiet click on the line, but it's not the sort of noise you'd notice unless you were waiting for it. And no one at Comsec has any reason to believe that calls are being tapped.

John starts to eavesdrop routinely. That was the way to find out what the enemy was up to, a way to anticipate the Texans' every move before it was made.

So here's John, listening in on Comsec's lines one afternoon when the security firm gets a call from a hacker named Craig Neidorf.

It is safe to say that no hacker was more famous than Neidorf in

1991. That was because Neidorf had beaten the feds at their own game a year earlier. In the months since, his legal fight had become legend. For years to come, wary prosecutors who were considering indicting hackers would caution one another to make sure their cases were airtight, so they could avoid "pulling another Neidorf."

The co-editor of the electronic magazine *Phrack*, Neidorf went on trial in Illinois in the summer of 1990, charged with fraud. The alleged crime was possessing and publishing an allegedly proprietary phone company document in an issue of *Phrack*. The government argued that the information was worth thousands of dollars, based on estimates from the phone company. But midway through the trial, the defense showed that the document's so-called proprietary information was publicly available; Bellcore sold the information to anyone who had thirteen dollars to pay for a technical article. Humiliated, the federal prosecutors in Chicago dropped the charges before the case reached a jury.

Yes, Neidorf was a hero to some hackers. But his notoriety also made him a target for anyone in the underground determined to make a name for himself in cyberspace.

Now, in the middle of a workday, Chris Goggans has answered the phone at Comsec's end of the call and Neidorf is on the other end. (John eavesdrops noiselessly; he's so quiet that he isn't even *breathing* as loudly as he normally would.) The phone call is just a friendly chat, but today Neidorf is frankly annoyed.

The problem is that anonymous callers have been phoning Neidorf at home and harassing him over the line. He doesn't know who is responsible, but he wants the prank calls to stop. They're really annoying.

"Sounds like they're doing stuff along similar lines as what they're doing to us," Chris says.

"Someone just called up my dad's house in Virginia," Neidorf says.

Chris is not surprised. He's outraged on behalf of this potential client, of course, but definitely not surprised. He even has a theory about who might be behind the calls.

"Sounds like Corrupt," Chris says, recounting his suspicions that John Lee has also been pranking him in Houston. "It sounds like something he would do."

At that moment a second phone line rings in Houston, another incoming call for Comsec. Chris asks Neidorf to hold on a minute, then answers the other line.

The voice on the second phone line says to Chris, "Yeah, that does sound like something I would do."

John couldn't resist. This was just so good.

In an instant, the reality hits Chris. He can barely believe it. The hackers are bugging the hacker trackers! And then calling up to boast about it! Impossible. But true.

Chris hangs up on John, gets back on the other line with Neidorf, and barks tersely, "They're listening to our call."

When he gets off the phone, Chris is so mad he can't think straight. John Lee has been eavesdropping! On Comsec's private phone calls! For how long? How often? What has he heard? What has he told his little friends up there in MOD? If this got out, Comsec would be a laughingstock!

Would you hire a computer security company that couldn't keep its own phone lines secure?

Chris calls the FBI.

He leaves the Comsec office, goes to a pay phone—at least this line is secure—and calls an agent in Washington, a name Craig Neidorf gave him, and tells the FBI agent that he has proof that hackers have been illegally listening in on his phone calls.

Of course, Chris has no idea that the Secret Service and New York Telephone have been keeping tabs on every electronic move that John Lee and Julio Fernandez make. He has no idea that an investigation is already under way. Chris has no idea that Secret Service Agent Rick Harris would quickly learn from his fibbie cohorts of the Texans' "cooperation" and view Comsec's involvement as a "nuisance." It would not help the government get convictions if the U.S. Attorney's office had to explain to a jury that the Southwestern Bell intrusions are nothing more than a pissing match among a bunch of kids.

Chris figures he's a businessman and has a legitimate right to protection. He tells the FBI in Washington that Comsec can't have hackers listening to calls from clients who are trying to outwit hackers themselves.

The FBI agent says he is very concerned.

A few weeks later, Chris and Scott go to the FBI's Houston office to discuss the full details of their beleaguered situation. They name names. MOD names. They tell the agents that MOD has access to Southwestern Bell switches.

Chris thought the agents looked "rather shocked."

But when the agents reported back to the office, they learned that a lot of people already knew about this electronic gang war. In fact, the FBI and the Secret Service were wrangling for control over the investigation of the case against John and Julio. Suddenly, everyone seemed to know what a switch is. Representatives from the U.S. Justice Department's newly created computer crimes unit traveled from Washington to New York City to be briefed. The FBI wanted to run the show. But the Secret Service was already in place. They're rivals. It got heated. The subtext was clear. This was going to be a big case, and everybody wanted a piece of it. This case could be a ground breaker.

Why did the case suddenly interest even the highest levels of federal law enforcement? Why did this pattern of intrusion, no different really from what the MOD boys had been doing to New York Telephone computers two years ago, now capture the imaginations of the best minds at the U.S. Justice Department?

Suddenly, it had dawned on everyone that this was a whole new area of crime. This was the future in law enforcement. It was going to explode and they had to get ready. This case didn't seem to be only about hackers and the coming information highway. It was about any criminal out there who would have the wherewithal to use high-tech, sophisticated communications to try to outwit the government and commit crimes. This case became a *first,* the first ever of its kind, and it would set the precedent for years to come in how the government tracked techno-literate criminals.

For the first time in U.S. history, the authorities wanted to put wiretaps on computers. This wouldn't be called a wiretap—no, it would be called a *datatap.* They wanted to identify the boys' every keystroke. Of course, for starters the government also wanted to put traditional *wiretaps* on John's and Julio's phones.

With evidence from the DNRs, from the traces Kaiser did in the spring, and from phone toll records that revealed every number John and Julio dialed, a judge was convinced to authorize the wiretaps—and the datataps. The Justice Department in Washington, D.C., decided that the Secret Service would run the show, setting up a monitoring headquarters called a wire room. The FBI would assist. That was a big victory for Agent Rick Harris.

Then reality hit everyone. Getting a federal judge's permission was the *easy* part of this task.

In a normal case, the Secret Service might need a modest amount of equipment to conduct a wiretap on a phone: a few tape recorders, some headphones, a couple of agents working twelve-hour shifts. Oh, and a notepad for the agents to jot down interesting things they hear and want to remember.

This was no normal case.

To set up the simultaneous phone-modem taps in the MOD case in 1991, the Secret Service first obtained an entire suite of rooms in the World Trade Center, where its New York City office was already headquartered.

One-of-a-kind, high-fidelity digital data intercepting equipment was brought up from Washington. More was designed and built on site for the case. A guy with a soldering iron and wire clippers worked away on the stacks of machinery that challenged the strength of the office-issue tables that had to support the weight. The bank of equipment was fifteen feet long, a massive, electric-shop mess of cords and wires and blinking lights, audio components, and big disk drives to store information. Shelves of dinner-plate-size storage tapes lined the walls, and a computer anchored it all, keeping track of the kind of information a DNR could accumulate: number dialed, duration of call, time and day.

The wiretaps on the phones were easy to operate. The government had been doing it for years, and three copies of every phone call were simultaneously recorded so that the court could seal one copy as insurance against tampering, the FBI could seal a second copy for its archives, and the prosecutors could use the third copy to build their case. This configuration of contraptions and tapes, and copies of tapes, and copies of copies was confusing enough, what with the government monitoring two phones at once. Three copies for Julio's calls. Three copies for John's. But the datataps—now that took some doing. You were trying to intercept data from two different computers talking to each other and expecting no interference, and you were not a hundred percent sure about the configuration of those computers. Each computer had its own, unique modem, and modems are quirky.

You've got to insert another device into the middle of this intensive data transaction. When the modem on John's computer meets another modem, it sends out signals. Hi, this is who I am, this is my baud rate, this is how I send information, let's shake hands, pleased to meet you, *beeeeppp*. If you want to hear the modems' conversation, you've got to use some intercept that will do the job without interfering with the idiosyncratic and sensitive signals. It's like breaking the infrared beam of a burglar alarm without tripping the system. Don't mind me, I'm just standing here quietly. Good luck.

It took a few days to get the datataps running properly, a few more days to capture a reliable stream of data without causing the modems to crash. Then it took a few days to tune the datataps to pick up information that actually meant something and didn't just appear to be a string of random gibberish. But a little welding here, some clipping of wires there, and the federal government's fledgling foray into data entrapment was under way.

Then, John got a new modem.

Everything crashed.

Out came the soldering irons.

There was a lot to keep up with, because the boys were supplying an enormous amount of phone and data conversations to eavesdrop on. Dozens of calls a day. The output was phenomenal, and the authorities were going through storage disks like so many dishes at a diner. Tom Kaiser went downtown for a consultation one day, and what stuck in his mind was the sheer number of people milling around the place—the teams and teams of agents that the government had assigned to pick up Tom and Fred's original investigation. Rick Harris had to work twenty-hour work days to supervise the government operation, but since he came from a long line of volunteer firefighters, he jumped to the call. It was in his genes—when the alarm goes off, he slides down the pole.

Under Harris's direction, two dozen agents worked twelve-hour shifts for weeks, listening to the intercepts, analyzing the information, transcribing it, keeping up a steady flow of the day's output so that

Prosecutor Fishbein would have the ammunition he needed to convince the judge to keep the whole operation up and running for a while longer. It was a wildly expensive undertaking. Every ten days, Fishbein had to go back to court, show the judge examples of the information being culled, convince him that the government taps were obtaining valuable evidence of illegal activities that outweighed John's and Julio's rights to privacy on their own phones. And their own computers.

Amid all this activity, Fishbein and Harris held daily meetings. In the morning. At supper. After midnight. They'd talk about the bits and pieces of information they were gathering; they'd get a judge's permission to send transcripts to computer experts for analysis. They began to realize just how widespread the intrusions were, just how knowledgable John and Julio were about computers. And Fishbein saw other names showing up in the transcripts, names of people that John and Julio talked to all the time: Mark Abene and Eli Ladopoulos. When he asked Harris about the names, Harris seemed to know who these boys were—

But by then, the sheer mass of voice and data interceptions was overloading the team. They could hardly keep up. But they had to keep up, because they were committed now—in up to their elbows. They were behind schedule in numbering and cataloging the data sessions, and the government wouldn't know what actually happened during those sessions until somebody listened to them. What were the kids saying on the phone *now,* anyway? What were John and Julio up to? What if John and Julio crashed a switching center that controlled the entire state of Texas? What if the Secret Service were monitoring during the crash, but didn't get a chance to listen until three days later? How would that look?

Harris and Fishbein didn't get much sleep.

Down in Texas, the informers had made an arrangement.

Chris had worked out a deal with the FBI that enabled him to tip off the feds anytime the MOD boys were active. If the New Yorkers called Comsec, for instance, Chris immediately left his office, walked down the street to a certain pay phone, and called Washington. The pay phone seemed secure, but the FBI agent has told Chris never to mention the agent's name on the line. Just to be careful. Chris thought the agent was leery that the MOD boys might have bugged the phone somehow and would prank the agent at home if they knew his name ("Mr. FBI Elite . . . Mr. FBI Elite . . .").

The Comsec crew didn't stop there, though. Comsec was a computer security firm, so it was Comsec's job to inform other companies if their computer systems were insecure, right? Might not be a bad way to drum up new clients, come to think of it.

In the fall of 1991, Scott Chasin phoned Information America, the company that collects tidbits of personal information about people's employment histories and finances. Scott introduced himself to the company's security department, then said, "Do a search in your system for my name. Do you see all the credit reports pulled under my name?" The Information America official searched and found ten recent requests for Scott's credit reports. In other words, ten people had logged in to inquire about Scott. That was pretty unusual, because an individual's credit report usually gets pulled only if the individual has invited scrutiny by applying for a mortgage or more credit cards. Some people's reports don't get pulled ten times in their lifetimes.

"You guys have got to do something about this," Scott said, explaining that he suspected MOD members had been fiddling with his credit information in an attempt to harass him.

The security official at Information America was astounded. How did they get into our system?

Easy, Scott said. The same way they've been breaking into TRW, Southwestern Bell, and the Bank of America.

They've broken into the Bank of America?

Oh yeah, Scott said. We've been watching them.

Scott said he had evidence the MOD boys had infiltrated a worldwide network that computers use to pass information to one another. He said MOD had access, through the network, to the private transactions of dozens of the world's largest corporations. MOD not only had access to delicate corporate secrets, but it also had the ability to upset a billion-dollar company's financial stability with a single keystroke.

And how did they do it?

Scott explained it with one word: Tymnet.

THIRTEEN

Parmaster had this special password.

Nobody knew where he'd got it, or when. It was just part of the Parmaster myth. But man, the things he could do with that password.

Parmaster's real name is Jason Snitker. He only happens to be in New York now because he's on the lam—no lie, he's actually being pursued by the Secret Service. Back in 1988, Jason was charged in Monterey, California—where he attended high school, read way too many spy thrillers, and discovered Jolt cola prior to becoming a fugitive—with breaking into a Citibank computer. The particular computer he was charged with breaching was one that assigned debit card numbers to customers of a Saudi Arabian bank. While he was in the computer, Jason decided to take a souvenir—approximately ten thousand valid debit card numbers. Some credit line.

Almost immediately, those card numbers started to turn up on bulletin boards from California to Texas. Free money! Hackers all over America started charging their long-distance calls to Saudi Arabian debit cards. It was like Robin Hood was sharing the spoils with his Merry Men. And by the time Citibank caught on, who knew how much money had been lost? When they finally caught up with Jason, the bank was asking for $3 million in restitution. Good luck.

Jason turned himself in on December 12, 1988, and that might have been the end of it, since he was a juvenile. Community service, don't do it again, blah blah, you know the routine. But Jason was an impressionable teenager. Remember the spy novels. He believed he *knew too much*. Jason saw a lot of secrets in cyberspace during his hacking days, including what he said were government secrets, like the Killer Satellite. That's how people say Jason described it—the Killer Satellite. According to documents he found on TRW's internal computer network (in addition to being a national credit-reporting agency, TRW is a giant defense contractor), the Killer Satellite, once deployed, could focus a death ray on other spy satellites, taking them out of commission. This was Top Top Secret stuff, Jason told people. It made the debit card numbers look lame.

Maybe the government would *liquidate* Jason to keep him quiet. Maybe it was all a big conspiracy, and he was the only one who knew the truth. He could really get hurt. So Jason freaked. No way was he going to wait around for jail time. Did Lee Harvey Oswald ever see jail time?

So Jason split, and now he's in New York, hiding out around Coney Island. The Secret Service have gotten a tip he's in the area. Ditto the New York State Police. One day, a state police senior investigator goes to a *2600* meeting at Citicorp and insinuates himself among the potted ferns on the balcony. The investigator is looking for Jason. He even brings a camera and takes surveillance shots. But Parmaster doesn't show up.

Some other people have found Jason, though. He's hooked up with MOD.

It's only natural that the best hacker on the West Coast should hang with the best hackers on the East Coast. It turns out they have a lot in common and go way back. Jason can't even remember the first time he encountered Mark. He knows it was on Altos. All Mark cared about then was hacking the phone system. That was cool with Jason. To each his own.

Now, though, the MOD guys had mastered the phone system and were looking for new terrain to conquer. Tymnet had potential.

And Jason had the password. One day, he shares it with Mark during a trade.

That's how it starts.

* * *

The password got Mark, John, and Julio into the heart of Tymnet.

In the past, they'd had limited access to this global electronic data network. For years, the MOD boys had been collecting Tymnet dialups, connecting to its system, then leapfrogging into other states or countries. It was kind of like long-distance telephone service for computers. Free.

In fact, while on the run, Jason had created a special account that would allow anyone, anywhere, to use Tymnet free. You'd get your computer to call a local dialup. Then the computer would ask you for a login, and you'd type "PARMASTERX75" (and then a password—parmaster=tymnet god—to prove you really were the authorized user).

Too cool. Word spread throughout the underground. It was operable for a whole year, enhancing Parmaster's image more than a PR firm could.

Although the Parmaster Network User ID enabled the MOD boys to use Tymnet, it kept them from tinkering with its engine, kept them out of the guts of this enormous repository of information and power. But with Parmaster's new privileged password, the MOD boys could explore dozens of dedicated subnetworks that the system links.

Tymnet's customers are big corporations and government agencies with far-flung computers that need to talk to one another. Now the MOD boys could get into Tymnet just like any engineer at Martin-Marietta Missile group could. Just like a vice president at Bank of America could. Just like a general in the U.S. Air Force could.

One thing separated the MOD boys from those other users, though. Most Tymnet customers log in to only a tiny section of the system, into the subnetwork that houses their organization's business. The MOD boys were different kinds of customers. They had the power to run rampant through *everybody's* networks.

Not right away, of course. At first, the MOD boys were simply blown away by the immense possibilities that lay before them. Sure, they'd hacked other big systems. They'd mastered the nation's entire phone system, hadn't they? But Tymnet's sphere of influence and importance is much broader than the phone system's.

Tymnet exists at the very vortex of cyberspace. Billion-dollar financial transactions, top-secret plans for fighter jets, the confidential credit history of the President of the United States. It's all there, sucked into cyberspace and rushing past at an incalculable rate. Information rockets through the pipeline that Tymnet controls, information packaged as precise units of data that zip from one user's computer screen to another halfway around the world. If you know the right commands, if you execute them at the right instant in time, then you can examine that data as it zips past. You can X-ray the units. You can become an electronic voyeur. You can, quite simply, know all the world's business.

Jason's password was the first step.

It is Mark, of course, who figures out the next step.

He logs into Tymnet day after day, obsessively, unable to take a break to sleep or eat, fueled by cigarettes and Coca-Cola. He sees a nearly infinite system, a system that is beautiful because of its endless complexity, a system that defies explanation, spiraling forever outward, like the Big Bang, always changing as new software and new hardware come online. He sees a system that he alone might understand.

Who knows more about so many different networks? AT&T employees only know about AT&T. New York Telephone engineers only know about New York Telephone. A Tymnet technician is just as myopic. But Mark sees how all the systems work and how they all fit together into a universe that he alone has conquered. Mark synthesizes that information into a worldview that allows him to go further, intuitively, hacking. That's why he never doubts that he will master Tymnet.

One day, Mark finds a few old computers called PDP10s, old minicomputers owned by Tymnet and used by the company's technical staff to store . . . administrative manuals! This is a find of unbelievable good luck. Imagine being an archaeologist, patiently digging through layers and layers of alluvium. And then you find the Rosetta stone. All the pieces are in front of him, but he doesn't know how to read it. Not yet.

John and Julio have told Mark they've heard that TRW operates its own subnetwork somewhere on Tymnet. It makes sense for TRW to exist on Tymnet, because car dealers and loan officers and bill collectors from all over the country had to get into the TRW system quickly to look up credit histories. All those TRW customers wanted a local phone number to dial to get in. Tymnet, with dialups in every major city in America, was the conduit.

John and Julio give him a number of valid users' names—who knew where the names came from, maybe a garbage dumpster, but they don't know what to do with the stuff.

But Mark does. He logs in, then types "TRWNET," and there he is. He's in the subnetwork. He gets a prompt:

    TRWNET>

He starts testing certain ubiquitous commands. He types "dir," and sure enough, a directory of file names appears. He types "type [filename]" and the contents of the file in question fill his laptop's screen.

Mark discovers that TRW has its own PDP10 computer on Tymnet. He starts looking around and finds a list of every account name and password. The subnetwork has files, and directories, and packets of information that you can send back and forth between users' computers. The subnetwork uses exactly the same hardware as the rest of Tymnet. If Mark can understand the architecture of TRWnet, by extension he'll know everything there is to know about the whole Tymnet system.

So that's exactly what he does. He breaks TRWnet down to its component parts and figures out how each one works alone, then how they work together. He learns the lyrical acronym for Tymnet's whole gorgeous architecture—ISIS, the Internally Switched Interface System. ISIS is made up of nodes, which are really just computer-processing chips. Each node is loaded with "slots," which run software applications. For instance, every slot has software used for troubleshooting or debugging. The software is called the Dynamic Debugging Tool, or DDT, for short. Get it? A debugger called DDT. Computer people have the best sense of humor.

He learns that there are supervisor nodes. They are to Tymnet what switches are to New York Telephone. Supervisor nodes are like traffic cops, routing packets of data to their destinations and keeping track of what's going on in the rest of the network. With the manuals he got from the PDPs, Mark has a blueprint, and he learns about the software tools that Tymnet technicians use to keep each node running. It's a great system because each engine he explores is loaded with its own full set of software tools, like DDT, right there in the slots.

One of those tools is the aptly named X-RAY, which allows Tymnet workers (and now Mark) to look inside each packet of data as it whizzes by. He learns about every aspect of the network, just as he did with the telephone system. And there's no stopping him. He hits Tymnet at least ten times a day, logging what the government will later say was an astronomical 23,314 minutes of network time.

Mark gets a stack of floppy disks that he calls "The Main 10."

He has copied all the information from the Tymnet manuals onto the floppies and carries them around. He's amazed at his treasure, amazed that there are other engineers out there in the world who think just like him, who so meticulously write down all this technical stuff for others to read. He's the only one with the Big Picture. Of course, he shares much of what he knows with his friends. It's only fair to give something back to Julio. And doesn't Julio share what Mark's told him with John? They're all friends. All members of the Masters of Deception. John is as compulsive as Mark when it comes to Tymnet—he can't resist logging in every day. The only thing that slows him down is the bad power supply in his building. Sometimes John's computer blows a fuse.

Tymnet has become MOD's new playground. The hackers are just like a bunch of excited four-year-olds, running around and trying all the rides at once.

Every day, it seemed, the MOD boys were finding a new expanse of the network to explore. One day, someone, it might have been John, found this weird list of subnets stored on one of the PDP10s. The list says Honeywell, Northrop, Loading Dock.

It sounded funny. Honeywell and Northrop, they were in the newspapers. Big businesses, right? Something to do with the . . . defense industry? But Loading Dock? What the hell was that? Some kind of multinational shipping concern? Or perhaps a purveyor of pressurized bulkheads?

There was one way to find out. The PDP10 tells them how to get into the Loading Dock host: Log in as COLORS. The password is "RAINBOW." How can they resist?

The MOD boys connect. But instead of seeing verification that they've entered Loading Dock, they see something very different on their screens. The host they've cracked is called Dockmaster.

They learn that Dockmaster is maintained by the National Computer Security Center, under the jurisdiction of the National Security Administration. The NSA makes the CIA look like lightweights, as any afficionado of spy books could tell you. The agency is so top secret that until recently, the government didn't even admit it existed. It's well known that the NSA earns its keep by monitoring the communications of the world, from telephone taps to satellite traps.

And here was a bunch of kids in Queens spying on them.

Anyone could see they were getting into *War Games* territory here. But the MOD boys venture further. The authorized user who owns the "COLORS" account happens to be an Air Force general, which the boys learn, well, by reading his electronic mail. The general is in a Dockmaster access group called AF, which they figure stands for Air Force. They snoop around, and to tell the truth, it gets boring. The general doesn't have access to anything that interests them.

In fact, the whole experience gives Mark the creeps. He calls it "spook stuff," and vows to stay away. Besides, he's much more interested in the information *pipelines*—how information flows through the Tymnet system—than he is in the information itself. That's one big difference between Mark and John.

John likes to sit and watch packets of information as they zip through Tymnet. He has started to research the names of the companies listed on the PDP10s to see what kinds of business they do and whether their transactions are worth more than a look. It was easy for him to dismiss Chiquita Banana or Florsheim Shoe Co. But there were dozens of company names he has to look up in Dun and Bradstreet.

One morning on the phone, while a Secret Service agent surreptitiously listens in, John tells Julio about a whole list of Tymnet customers that he found while scanning the network the previous night. He says he found a thousand new computers, belonging to such customers as Sealand Services, Black and Decker, Exxon, the

Department of Motor Vehicles in California, and the administrative
offices for the U.S. Court system.

John also says he's found computers that belong to Randolph Air
Force Base.

"Oh God," says Julio. "We've just got to start hitting those sites
left and right."

"That's what I plan on doing," says John.

John, Julio, Allen, and Eli have become fascinated by the possibilities of TRWnet for reasons that are different from Mark's. Now,
instead of possessing a couple of TRW accounts culled from social
engineering or bulletin boards, a couple of accounts that could be
discontinued at any given moment, suddenly *MOD had the whole list
of accounts.* It was getting pretty intense, all this power. They could
look at whoever they want—a guy who Eli says his dad is thinking of
suing, a guy in Brooklyn who has a car Eli wants to buy—and find out
confidential facts.

It was unfortunate that simply having the authority to change
credit reports didn't give them the necessary knowledge to actually
do it. They'd been thinking that Mark might have figured out how to
change TRW credit information. You know, clear up a bad credit
history. Or make someone's good credit background lousy.

If Mark had that knowledge, he wasn't saying.

One day on the phone Allen and John talk about whether another
friend, named Matt, has gotten a TRW manual to help them.

"Did you talk to Matt? Did you ever find out how to put the
delinquencies on?" asks Allen.

"He forgot to bring it," says John. "I'm really going to get on him
about that. . . . He has the complete instructions to do anything."

"Oh yeah?"

"Yeah," says John. "I'm talking about everything. . . . He didn't
have the bullshit thing that the customers get. He has the admin
manual, and he keeps forgetting. It says how to remove them, how to
add them, how to do all types of comments to destroy peoples' lives
or fix them. You know, make them look like saints."

"Yeah."

Destroy people's lives? Make them look like saints? Is this what
hackers do?

*   *   *

When Mark was not around, the less scientific members of the MOD family still liked to look up celebrities. They collected celebrity credit histories like baseball cards. Now it was so much easier than it once was.

One Tuesday night, at 10:21 P.M., Eli phones Julio.

"So I was looking up like a whole bunch of famous people," Julio says.

"Yeah?" Eli says. "What'd you get? Anybody good?"

"I got, uh, who's that? Richard . . . Gere."

"Does he have anything good?" asks Eli.

"Well, I mean they all have good stuff. I didn't get it to get their [credit] cards, you know."

"I know," says Eli.

"I have Tony Randall," says Julio.

"You ever get a social security number on that?"

"Yeah," says Julio. "But like, I'm always at a loss with who to look up, you know? I looked up Julia Roberts, but I don't know if she has an apartment in New York City, because I came up with like a million of them."

"How'd you get Richard Gere's address?"

"Because—well, it was obviously him because he was, like, employed by Paramount Pictures or something."

Sometimes the MOD boys lurked silently in Tymnet's administrative accounts, watching technicians even at the moment when they changed the daily passwords. Lots of times they staked out the network and would know the password of the day before some nitz of a technician, and then they'd watch him as he tried to use yesterday's discontinued login. Duh.

Mark, Julio, and John could read the Tymnet security department's e-mail, and so they could scan highly sensitive memoranda, bulletins, and alerts about Tymnet security. That way, they always stayed a step ahead. They knew about any new security devices well before the devices were put into use.

Of course, none of this was happening in a vacuum. Tymnet was well aware of intrusions into the system, and had been ever since the day months ago when Southwestern Bell noticed John and Julio on their switches. Tymnet knew that intrusions into its system were not uncommon. Tymnet was biding its time.

Even if Tymnet were ignorant of the intrusions, Chris and Scott called security administrators every day. The Comsec crew knew all about MOD's doings, even got a glimpse of where they were going when they raided MODNET and found some how-to-hack-Tymnet files.

"These guys are all over you," he says. The Tymnet guy seems very interested.

Chris says they can work out a deal. "Well, you know, you guys are so wide open. We're going to have to go to our clients and say, 'Your links to Tymnet are a weakness.'"

The Tymnet official tries to dissuade Chris from making those calls, and promises that Tymnet is already working to patch the holes.

"Oh no, you're not," says Chris. "We know exactly how weak you are and here's proof." And then he faxes the Tymnet guy some of the files Comsec got from MODNET.

That convinces the Tymnet official to make a deal. In exchange for Comsec's silence, he offers Chris and Scott a user ID to make free phone calls through Tymnet.

The Texans accept it and use the ID to call European chat systems where they gab with hackers. That was part of their job, gathering information on how hackers work.

The network user ID is t.cds01, which stands for Tymnet Comsec data security number 1. Way cool.

Chris and Scott also devoted a lot of time to tracking MOD's forays through Tymnet—hey, it's a living—and called the Tymnet official pretty often just to say, look what MOD is doing now. Comsec was determined to crush the MOD boys. Any way possible. They wanted to hurt them. It was all about business, they told people. It was bad for business whenever MOD made Comsec look silly.

Like the time when John posted Scott Chasin's mom's credit history where everyone could read it, right on an electronic bulletin board that thousands of hackers log in to. The credit report listed Scott's mom's home phone number and address, as well as the phone numbers and addresses of some of her neighbors. John also posted some sexual comments about Scott's mom, for good measure.

Right at the top of the record, it said that the information came from Information America, a system the MOD boys reached through Tymnet.

Scott didn't like to think about what his mom would say if she ever heard about this incident.

Chris didn't like to think about what Comsec's clients would say. It was definitely bad for business, Chris said.

More calls ensued from Texas to Tymnet.

The Texans were starting to make pests of themselves, though, and the Tymnet security department was starting to dread the calls. Wouldn't you? It was never *good* news.

Soon the Texans were connected with another security officer, a senior investigator with Tymnet named Dale Drew.

Drew had once run a bulletin board of his own, using the handle The Dictator. The Dictator had been an underground informant in Operation Sundevil. All the MOD boys had heard the rumors in the underground that Tymnet was hiring The Dictator because of his intimate knowledge of hackers' habits.

In this case, however, he knew whenever the MOD boys were in his system simply because they taunted him. They couldn't help it. Maybe because he used to run a BBS, they thought he was an ex-hacker. Maybe they thought that made him a player in a game called Computer Intrusion. Maybe because The Dictator was an underground informant for Operation Sundevil they declared a free-for-all. Some game. In any case, Dale Drew would win if he kept the MOD boys out. They'd win if they got in. All good fun.

The MOD boys believed that Dale Drew was logging in to electronic bulletin boards anonymously, trying to insinuate himself into MOD, trying to get information about just how much the hackers knew about how Tymnet works. In any event, the MOD boys were retaliating by telling Dale Drew that they could read his electronic mail. They told him they could read his memos to his bosses. They even told him they could read the new passwords he used to try to secure the Tymnet system from them. They were trying to drive him crazy.

The MOD boys believed that Dale Drew, in return, was gunning for them. They believed that Dale Drew was creating new passwords for the Tymnet system that referred to MOD specifically. One morning on the phone, John and Julio talked about the Dale Drew situation:

"What a moron," says Julio.

John asks Julio to give him Dale Drew's password.

Julio says the password has been changed, so he goes to look it up. Julio tells John he knows at least one new Tymnet password (though not necessarily Drew's password). The password is "P-H-I-B-star-S-U-X exclamation point," Julio reads off.

PHIB*SUX!

John thinks about this.

"Phibersux," he says.

"Yeah," Julio says.

"Phibersux."

"Yeah."

John can't believe it. Phibersux. It sure sounded to him like Tymnet was playing along. What a strange game.

FOURTEEN

It is the *freshest* thing Julio has ever seen, and he can't take his eyes off the TV screen. What did the announcer call it? A skycar.

For a moment, Julio loses himself in its two-seated splendor. For an instant, he's no longer a kid sitting in front of cable TV in the Bronx, watching an episode of "Beyond Tomorrow" on the Discovery Channel. The California-born skycar has just been unveiled at the Essen Motor Show in Germany, after twenty-five years of development. Julio imagines himself in the pilot's seat as the skycar zips past on the screen. For an instant, he doesn't hear the mice in the kitchen. He only hears the roar of the engines as the skycar lifts off vertically, like a rocket. He sees himself, twenty thousand feet in the air, guiding the two-seat car-size airplane wherever he wants to go. And Julio wanted to go a lot of places. He'd been to Peru with his mother, which was pretty fresh, too.

But can you imagine the possibilities? Anywhere you wanted to go. It was like hacking into Tymnet, punching a couple of numbers, a couple of letters, and you got to go to a computer in Canada or something. You were in Canada. Or were you? It didn't seem that way to Julio. You were just in another computer. You were still in the Bronx. But in a skycar, you could really go to Canada if you wanted. And when you got there, you could slowly lower it down into a tight parking space.

And it would be so awesome.

Yeah, right. Where was Julio going to get the money?

He didn't even have the money to buy a damn pair of shoes. His mom was out of work, so she wasn't buying him any, either. Julio's eighteenth birthday was coming up this week, but he didn't expect any presents, let alone a skycar.

It was funny how things work. Here you were, obsessing about a tiny airplane that would take you to the stars, wondering how you'd ever get that kind of money, when the phone rings. And opportunity comes looking for you.

The MOD cowboys have ridden where no one else has ever gone before in cyberspace, and they've brought back tales of the riches they've seen. A lot of people could put MOD's knowledge to profitable use. It was only natural, of course, that word got around. And it was only natural that people would want to pay to learn how to get some of those riches for themselves.

Of course, offering a hacker money for information was a rather *delicate* proposition. It was one thing to joke around about selling computer information. It was another thing altogether to actually do it.

For starters, you had to be sure you were making the offer to a receptive audience. No way would anybody offer Mark a penny for anything. He would look at you as if you were some kind of moron. He didn't *sell* what he knew. If he liked you, he told you what he knew, free. If Mark didn't like you, well, he'd let you know it.

And Paul was no good. He hadn't even been hacking into Tymnet, and even if he had been, he wouldn't sell information, either, because that would contradict the hacker ethic that he believes so fervently. And Eli? He didn't know enough to sell anything. Besides, he'd never replaced his computer after the raids.

No, if you wanted to approach the Masters of Deception, you had to do it carefully. You had to make the right move on the right guys. On the hungry guys. If you wanted to approach MOD, use a middleman who was familiar with the players, somebody who knew the situation.

Somebody like Alfredo.

And some guy named Morty has been persistently bugging Alfredo, calling him all the time, asking him to broker the deal. Morty has money.

Alfredo knew who to approach. He knew, for instance, that Julio had been telling people that he was pulling confidential credit information for a private investigator. Who knew if that was true? But the fact that Julio was saying it said something about him, didn't it?

Alfredo became Morty's go-between. It happened in late November of 1991. It was as easy as this:

Alfredo calls Julio at about two o'clock on a Monday afternoon. Julio's already on his other phone line, chatting with a friend about getting together later to hang out (no MODster does *homework* after school, so there's a lot of free time to fill). Julio puts his friend on hold to answer the phone.

"Julio," Alfredo says.

"Yeah?"

"Anyway, listen."

"What?"

"This Morty, yo man, I went to his house; he's fucking punk," Alfredo says. "This guy got money."

"Yeah, I know. Parmaster already told me all about this guy," Julio says.

"Oh yeah? What'd he say?"

"He said this guy is like, he's like one of you; he like goes into your money-making schemes," Julio says.

"Yeah, but they work."

"Yeah, I know. I mean, I already heard about it. Like Par already told me he could set me up with this guy to make like thousands. I was just, I was like, uh, whatever you say, because I don't go for that stuff." Cool-playing Julio.

"He wants to talk to you," Alfredo says. "He told me . . . let me talk to someone in MOD, but you can even be the middleman. You know what he wants to buy?"

"Yeah. I know exactly—plane tickets, right?"

"No."

"Oh? What?" Julio asks.

"He wants to buy an Information America account."

So here was the deal, then. Morty Rosenfeld needed an Information America account so that he could pull up credit histories. That way, he could get people's credit card numbers. What would he do with the credit card numbers? Morty wasn't saying.

Morty was no stranger to buying and selling credit information. In fact, the past summer the feds had charged Morty with selling a password to the credit-records company Credit Bureau Index to somebody who'd used it to obtain credit card numbers that were used to order electronics equipment.

Morty has been into the hacking scene for about five years; he'd gotten his start at age fifteen as a member of some gang called Force Hackers. Morty now has a crazy-ass plan to build and sell his own version of IBM personal computers. The Rosenfeld 1000SX! The ultimate hacker's box! All he needed to do was come up with a way to buy lots of computer parts, components that he would fit together to create the machines.

Morty lives over in Coney Island. In fact, Parmaster is staying with him. Well, actually, Parmaster *was* staying with him, until the authorities caught up with the fugitive sleeping on a fold-away bed in Morty's living room. Parmaster had been arrested about ten days ago. The feds told Parmaster they wanted to ask him some new questions. Over the past couple of months, TRW had reported unauthorized intrusions into credit files, and it seemed that the unknown intruder had routinely checked the credit report of Parmaster's mom.

Parmaster's arrest hasn't slowed down Morty, though.

Julio is intrigued by the possibilities of working with Morty. Maybe he was feeling sorry for himself about his birthday and no money. He doesn't let Alf know any of this, of course.

He puts Alf on hold. This is where the negotiations start to get pretty complicated, as Julio switches back and forth between Alfredo and his friend on the other phone line. This is how Julio firms up the deal. Watch him work. The feds do. He's good.

Julio switches over to the line with his friend, and says, "You won't believe who called me up.

"Alfredo," Julio confirms. "Yo, he got dough up the butt now. . . . You want to make some money today?"

"How?" asks his friend.

"Remember those Info Am accounts that I was pulling for that P.I. guy?"

"Yeah."

"Okay, this guy wants to buy one off of me for four hundred."

"Alfredo?"

"Not Alfredo. Come on," Julio says. Wake up.

"Tell him to shut the fuck up. And he'll give you four hundred dollars for one . . . ?" Julio's friend is incredulous.

"Yeah, for an Info America account," Julio says.

"Oh my God."

Julio puts his friend on hold and switches back to Alfredo.

Julio's playing the convince-me game: "I don't want to do anything with this guy."

"Oh, you don't?" asks Alfredo.

"No, I don't think so."

"Why not?"

"Because like, I don't like dealing with this, and like . . . I've already been offered it by like many other people, and I, you know, why do it with him when I could have done it with like a million other people?" says Julio.

"Well, you can do it through me," Alfredo says. "That's one good part."

"What's that? That's not good."

"No one can even see you. He won't even have your phone number," Alfredo says. "He won't know who it is that gave it to me or anything . . . I can probably get him to go up to five hundred."

Julio mulls this over, then says finally, "If he comes up here, then I'll think about it."

"I can make him go up there."

"Today, though?"

"Probably. I can have him drive me up," Alfredo says.

"Oh, he's got a car?"

"He has access to one."

"Hold on," Julio says.

Julio switches back to the other line, where his friend has been waiting patiently for news of the negotiations. Julio's way ahead of Alfredo on this one; he knows nobody is seriously planning to drive up to the Bronx to make the deal. To his friend, Julio says, "You want to go down there?"

"For what?"

"To get the dough."

"Serious?" his friend asks.

"Yeah. Hold on. Let me make sure he's got it today waiting for me. Hold on," Julio says.

Back to Alfredo.

"Could he have it there waiting for me today?" Julio asks.

"What, the money?"

No, a firm handshake, Alfredo. What do you think?

"I guess he could," Alfredo says. "I'll call, hold on, let me call him up," Alfredo says. "Let me go get his number—wait up."

"All right, I'll be back in a second," Julio says.

Alfredo is consigned to hold again.

Julio goes back to his friend, to whom he confides, "He thinks so. He's calling the guy up right now."

"Oh my God."

"That . . . guy is for real because I've known him through someone else as being like a big spender and everything, you know? Like Alfredo is full of bullshit. If he says five hundred, expect three hundred, you know?" Julio says.

"Yeah."

"Like, he's just full of moronic bull. But the other guy I know is for real," Julio says. "I've got to score big. If I tell you why, you'll laugh at me and say, 'You're a moron, Julio.'"

"What?"

"No," says Julio. It's too much to share his secret.

"What? Like bootleg alcohol?"

"Yeah, come on, you know I'm not into that kind of stuff," Julio says.

"So what?"

"I want to buy, uh, a skycar. Yeah." There, he's said it.

"Oh my God."

"Shut up, okay? It was on TV. Did you watch TV last night?" Julio asks.

"Yeah."

"You saw the Bundys, right?"

"Yeah."

"Well, while you were wasting your time watching the Bundys, I was watching 'Beyond Tomorrow' and they got this fresh—no, I'm not even going to explain it to you because you're going to make light of it," Julio says.

"Julio, I've seen it before."

"It's fresh, right? Yo, I'll give my life for that."

* * *

These conversations, of course, were being recorded by the Secret Service in the command center suite at the World Trade Center, and meticulously transcribed. This call, in fact, was known as Call No: Red 3662-3697. Isn't eavesdropping the most amazing thing? You can see why people do it.

Julio switches back to Alfredo, who hasn't yet reached Morty.

Julio proposes that he and his friend come down to *wait* for Morty at Alfredo's, but Alfredo says that's not such a good idea—his grandmother was coming over today.

"We can hang out outside at McDonald's, and then like walk into your house once in a while," Julio says.

So a deal is struck. They'll meet at Alfredo's apartment, then go over to the McDonald's on Columbus Avenue, the one on 91st Street, a couple of blocks away. Alfredo tells Julio that if this deal works out, Morty might become a steady customer. He tells Julio that if he can provide a steady supply of Information America accounts, Julio will earn up to five hundred dollars a week.

That's a lot of money. Julio could use a lot of money. Five hundred dollars a week is enough money to buy a lot of pairs of shoes. That was enough money to share.

After Julio disengages both his phone lines, he gets another call.

From John.

"Yo, meet me at Alfredo's house in like two hours, okay?" Julio says.

"Why?"

"Because me and my buddy, we're going down there. We're going to meet this guy Morty. We're going to take Alfredo out of the fucking loop. He's not going to be a part of it."

"What do you mean?" John asks.

"Because Alfredo is already getting too close with Morty and Morty gave him like a dial last night . . . for like a TRW account," Julio says.

"Get out of here."

". . . Morty . . . runs like a PI thing . . ."

"Oh yeah?"

"And he's going to buy," Julio says. "Just meet me down there in like two hours, okay? . . . I need money for tonight's party."

Julio says that John should bypass Alfredo's house and go directly to the second floor of McDonald's.

"Why?"

"Because I'm going to sort of accidentally run into you . . . and then you'll be part of it," Julio says.

Now John's a part of the deal. Hey, that was cool.

One thing about John and Julio. They both had this impression of this Morty guy. They both thought he was rich. He had to be rich: he "has access" to a car.

"Did he definitely give Alfredo a thousand dollars or something?" John asks.

"Yeah, he gave him money, yeah."

But a thousand?

"I think so."

"But you don't know," John says.

"Even if we don't get that, you know, who cares?" says Julio.

"I know."

"I'd do it for twenty," Julio says.

"I know. Right. Like I would take a hundred," John says. "Because that's like a pair of sneakers and shit."

There is something incongruous about a McDonald's in Manhattan. You don't expect to see the Golden Arches right there, amid all the honking taxicabs and roaring fire trucks and panhandlers. But there was this big neon sign on the outside of the building, touting the McDonald's "space station" inside, which makes the place seem even sillier than usual. McDonald's in Manhattan always smells greasier than McDonald's in the suburbs. Maybe that was because you can't throw open any windows to let out any of that refried air.

Today, all the action was taking place on the second floor of the two-story fast food restaurant. Morty buys the burgers. John acts as the lookout while everybody eats. Yeah, Alfredo is there, too.

The sun was already starting to go down, and it was cold. It was the kind of weather that makes you want to fill your stomach with a burger.

Julio has arrived at McDonald's with valid account information that he plans to sell Morty, but before they do the deal, John pulls him aside and talks Julio out of it. Give him "dead stuff," John says, meaning accounts that are no longer good. That'll get Morty off your back, John says.

Julio isn't so sure. He really wants the money.

Of course, Morty doesn't hear any of this byplay. Morty has these dark tight ringlets all over his head, hair so curly you can't tell how long it really is, and he keeps looking around, kind of nervous, just looking.

So then, Julio and John give Morty some accounts that they say he can use to pull credit histories. They tell him it's the good stuff.

Morty gives them some money, just pulls it out of his pocket in a big fat wad, and peels off a bunch of fifty-dollar bills. *Fifty-dollar bills.* He gives Julio ten of them, and John gets four.

John decides the deal isn't such a bad idea after all. Seven hundred bucks for outdated accounts. Sure sounded like a good deal.

The meeting lasts about half an hour.

The MOD boys walk outside and leave Morty in front of McDonald's. They head down to a neighborhood pizzeria, Alfredo trailing. They have a lot to talk about.

Morty walks off in the other direction. It will be interesting to see if these accounts work, he thinks.

Of course, maybe Morty wouldn't be losing anything if they don't work—John would later say the fifties were counterfeit.

In fact, Morty complained to Alfredo that the accounts didn't work. After a number of phone calls, disclaimers, recriminations, and posturing, Julio finally gave Morty a valid TRW account. Morty proceeded to use the account, diving into the credit union database as if he were on some kind of supermarket shopping spree. In less than a week, Morty had pulled 175 credit reports.

One night, around ten o'clock, Mark and Julio are on the phone, just talking about the things friends talk about. Julio's not thinking about the deal he did with Morty; he's not thinking about the possible repercussions. He's thinking about something much more existential, like what would you do if you found out you had cancer?

"I'd just, I'd tell it, the whole world, everything," Mark says.

"If I knew I was going to die in a year, I would still keep it to myself for like eight months, you know."

"Until you were like on your deathbed or something," Mark says.

"Yeah."

"Then you'd like make appearances on national TV and tell the world."

"No, I would rather like give it to a couple of people," Julio says.

"Why? You think you'd only give it to a couple of friends?" Mark asks. "If I was going to die? I'd want to piss off the phone company as much as I could. So I'd like tell the whole world everything."

"No, you know what I would probably do?"

"What?" Mark asks.

"Instead of telling everyone, I would like crash everything."

"Nah, the crashing everything isn't fun," Mark says. "Because they could always say, 'Oh it was equipment failure,' before they'd admit that a hacker did it."

"Say I'm dying," says Julio.

"Or do it while you're on TV," Mark says. "That's even better."

"Yeah, like Channel Four, right?"

"It's like, here, I'm about to knock out phone service in Texas, click, and then you go, like, *demonstrate*, you go and call Texas," Mark says. "And it's like, why would you do something like this? Oh, I'll be dead in a couple of weeks."

"Like say I'm taking out 911 in the South Bronx," Julio says.

"Yeah, right."

"And then, imagine, like a doctor runs in. I was wrong!—those were your dog's X-rays!" Julio says. "And like you're perfectly okay, and then you're like, oops."

"I know. And then like you have a heart attack and die."

"No, you don't have a heart attack. You get arrested," Julio says. "And like, why'd you do it?"

"No, but then you have a stroke on the way to jail," Mark says.

"I would probably kill myself if I knew I was going to get in that much trouble," Julio says.

FIFTEEN

Julio could only remember seeing his father cry twice. Just twice, in his whole life, once on the day of Julio's grandfather's funeral and a second time on December 6, 1991.

That was the day the Secret Service raided Julio's apartment, carting off every bit of computer equipment in the place, even confiscating the dark-colored backpack Julio carried to school every day.

For months, Julio had been feeling confident, thinking himself immune to the process of justice, as he called it. He'd seen plenty of hackers get raided and nothing bad had ever happened to them. Why should he worry? Mark never went to jail. Eli never went to jail. Paul never went to jail. In fact, in the nearly two years that have passed since the first federal raids of the MOD case, Eli and Paul had never even been charged with any crimes. To Julio, those first raids seemed silly, almost mythical. The 1990 raids his friends had endured seemed like—well, like a badge of *honor* more than anything else.

But then, the Secret Service agents showed up at Julio's house at 11 A.M., and they carted out his Apple IIc, his monitor, his modem, his manuals and documents, his records and notebooks, his floppies, his cassette tapes. Just like that. It was gone. All gone.

Julio was not at home at the time. His brother Louis was there, though, and he told the agents that Julio was on his way to school. More agents out on the street were already following Julio. They had followed him down from the Bronx to Manhattan, where they were watching him making a call from a pay phone at Forty-second and Park. Later, they took his backpack, and all the cassette tapes and the floppy disk inside it. And when Julio got home, he realized this was not a joke, not at all. His dad was *crying,* and there went Julio's bravado.

Of course, Julio was not the only one who got raided on this cold December day. Allen Wilson's two-story, single-family home on Moon Drive in Fallsington, Pennsylvania, was also included. In the search warrant application, the Secret Service told the judge that during a phone conversation Allen had told John of stealing electronic mail messages meant for Chris Goggans. A photo of a federal agent at Allen's house would show up in Allen's local paper soon after the raid.

John got raided, too. He lost his Zenith compact TV as well as his computer. Alfredo got raided, but you might have expected that. He'd been running his own call-sell operation, selling people long-distance service over some unwitting corporate switchboard. And Morty Rosenfeld got raided. Mark got raided—the only member of MOD to earn a second visit. He sat on the couch in his living room, listening to an agent read his Miranda rights. "You have the right to remain silent . . ." The little speech was printed on a business card. The agent flipped it, and Mark saw "NYNEX" written on the front. The agent asked Mark to initial the card to acknowledge the exchange.

The federal agents took boxes and boxes of computer equipment. Really, sometimes it seemed more like a job for a moving company than a crack law enforcement agency.

And then, for a few months, history repeated itself. Nobody heard another word from the government.

Stephen Fishbein sat in his office high above Lower Manhattan, trying to create a criminal case from a morass. He had too much information, and no sense of a pattern. He knows John and Julio have sold access codes to Morty—that's obviously illegal. But beyond that isolated incident? Fishbein had millions of bytes of data from datatap sessions, pages and pages of transcribed notes from nearly five thousand phone conversations, toll records, DNRs, and now, boxes and boxes full of floppy disks, hard drives, and notebooks seized in the raids. He knows the case is bigger than just John and Julio and TRW and Southwestern Bell. But he has no idea how things fit together. Where to start?

He reads. And one day, he comes across a single document that sets a new direction for the case: "The History of MOD." The History of MOD told the whole story of how five boys knew each other, the scope of their relationship. It was a starting point.

There was an organized group involved, they even had a name for themselves, MOD, and they dated back to 1989. And here was John and Julio coming into the gang as Corrupt and Outlaw and . . . wait a minute—*they were talking about how they got raided by the Secret Service once before!*

"The History of MOD," the tale that Eli had concocted so innocuously, had become the blueprint Fishbein used to understand who all these boys were, how they all worked together as a group. Hey, the guys from the old case were totally in with the new guys. In fact, the old guys were giving instructions to the new guys! The document even hinted at murky "fringe benefits" the boys got from hacking. *It said each person in the group has his own illegal specialty!*

Before anybody knew what was going on, Fishbein and Harris were riding out to the Eastern District to talk to the prosecutors there. And now Fishbein heard the whole story. He knows Mark is a repeat offender. He knows this *conspiracy* dates back several years. He learns that the Eastern District even sent off all the evidence seized in the 1990 raids to Bellcore, where the earlier crop of floppies and drives and notebooks had been scrupulously analyzed. Page after page of Mark's notebooks, crammed full of phone numbers, commands, user names, and doodles of a surfer catching a wave in the margin—all had been analyzed. Bellcore's work was so thorough that the prosecutors could use the notes to say, with authority, that a certain phone number he wrote on page 3 is to a switch. A user ID on page 11 was a valid login you could use to get into the switch. This command here on page 67 functioned on the switch. Page after page of hieroglyphics decoded. As for the doodling, well, the surfer seemed to be just a surfer.

Fishbein decided the best thing to do now was to separate the cases of Morty and Alfredo from the case of the MOD boys. That seemed to be a natural division; the involvement of Morty and

Alfredo was on the fringe of what really happened. It would be easier to get indictments and it wouldn't confuse the jurors. Sure enough, Morty was indicted. Alfredo was indicted. What was next?

In the spring of 1992, a grand jury began to hear the evidence that Fishbein has accumulated, sorted, digested, and turned into a story: "The History of MOD."

The letters arrive in June of 1992.

The letters come in long white envelopes, and look very stern and official. They are mailed to the MOD members who got raided in 1990 and 1991. Mark, Paul, Eli, John, and Julio all get them. "You are the target of a grand jury investigation," each letter says.

Paul remembers it as an odd sensation, opening his letter, reading it, processing the information the letter meant to convey. You are the target of a grand jury investigation. Paul hadn't really been a part of any MOD group activities for more than two years at this point, he'd dropped out of the scene after the 1990 raids. And yet here it was, the whole mess from his late adolescence rising up to haunt him as a young man. Was Paul even the same person two years later? He was an adult now, an adult being held accountable for a child's actions. Would the grand jury realize that? *The grand jury.* Just saying the words to himself was terrifying. There was a group of people impaneled (*impaneled*) in some court, hearing testimony about him. They were sitting there, all day, in a room, hearing *witnesses* describe all kinds of illegal things he'd supposedly done. Who were these witnesses? What were they saying? What could they know of Paul? What, exactly, did he do, during those long ago late-night hacking sessions? He tried to remember.

Imagine a roomful of strangers all listening intently to the most intimate details of your hacking forays, all trying to comprehend exactly what you did when, all relying on the *prosecutor* to make this confusing technical jargon understandable. Fewer than one family in three even owned a personal computer, fewer still had modems. Good luck.

Allen's a really helpful guy. Sure he is. Look at him making these annoying trips from Pennsylvania to Manhattan whenever the prosecutor called. Allen's family hired a lawyer to accompany him.

The other boys heard that his lawyer had put together a nice deal for Allen. That Allen was cooperating against his friends. The feds were grateful, the other MOD boys hear. Allen's information was helping build a stronger case against them.

Most of the boys didn't have the cash to hire their own attorneys. Paul, Mark, Eli, and John will get court-appointed lawyers. Mark got assigned a good lawyer, but he still worried privately even as he called the charges "nonsense." It bothered him more than you could see. He felt it in his stomach, the rat gnawing at his gut. Eli's lawyer complained to Fishbein that Eli was being unfairly lumped in with the rest of the boys; after all, Eli hadn't owned a computer for two years. Yes, Fishbein says, but he had Eli's voice, in late 1991, on the phone asking his friends to provide him with illegal credit information from TRW. Paul got a court-appointed lawyer named Marjorie Peerce, who told him one day that the assistant U.S. Attorney in charge of the case wanted to speak with him.

That was how Paul and Fishbein met for the first time.

They were not likely to hit it off. Paul is shy and sullen, wears his hair halfway down his back, cinched in a ponytail. Fishbein is aggressive and sure of himself, an alumnus both of Yale Law School and a Scarsdale childhood, with neat dark hair and simple round eyeglasses. They are not that far apart in age, born less than fifteen years apart, in fact, but they might as well have been from different centuries. Fishbein likes opera and ballet, and he retires at night to a studio apartment where he plays classical music on an upright piano positioned in the exact center of an Oriental rug. He does not have a computer in his home.

Fishbein has told Marjorie Peerce that if Paul wants leniency, he should tell the truth about exactly what he did and when he did it. Fishbein felt some sympathy for these boys, although he could not empathize with their actions. He is not a man who has ever contemplated breaking the law. But most of the other defendants he'd faced as a prosecutor have been more obviously hardened criminals, men who were selling drugs, or involved in drug conspiracies, or taking bribes at banks or savings and loans. The crimes of these boys were different. They were committed in the first flush of adolescent bad judgment, not in the cold avaricious world of drug dealing and sleazy bribery.

Fishbein's objective at this stage of the game was to get as much information as possible from any of the MOD boys willing to talk. He wanted to hear Paul's story.

So Marjorie Peerce and Paul go down to St. Andrews Plaza, which houses the U.S. Attorney's office in lower Manhattan. The relentlessly modular building is jammed into a too-small spot hard by venerable Saint Andrew's Church. Two hundred years ago, Saint Andrew's Church must have looked grand, but now it looks like an old barnacle ready to be scraped away by progress. Paul and his lawyer pass through a metal detector in the lobby, and wear sticky badges that proclaim their visitor status. They go upstairs to Fishbein's offices, and Paul is struck by how businesslike the whole situation is. Secret Service Agent Rick Harris is there, too. They all sit around a table in a conference room, and Fishbein isn't trying to intimidate Paul, he just lays out the facts. He tells Paul this is real serious business.

Marjorie Peerce has told Paul beforehand that it would be a good idea to express some remorse at this juncture in the meeting, so Paul just lowers his head, trying to look like a pussycat when you yell at it.

Then Paul tells Fishbein that he hasn't been involved in MOD's computer intrusions since the 1990 raids. He doesn't admit to Fishbein that he's been secretly logging in to New York University's computer system, and then dialing out from an account there to make free long-distance phone calls. He doesn't admit that he's been doing the same thing with the Hunter College system in Manhattan. And that will count against him.

Someone asks Paul to leave the room.

The rest of them stay behind, and a few minutes later, Marjorie Peerce comes out to tell Paul that the prosecutors won't budge. There's no way Paul can escape being indicted. She says that the only way to escape is to testify against his friends. (What Paul remembers hearing even years later when he imagines this incident, is the suggestion that he *rat*.)

Paul thinks Peerce's suggestion over. He's been puzzling over this for a while now, ever since the rumors started about how Allen was cooperating. Paul thinks about how he's in the middle of college, how jail could really screw up his life. Paul thinks about how dumb this whole situation is, no place for the valedictorian of Thomas Edison High School to be, about how his explorations on the Laurelton switch occurred so long ago that it feels like a different lifetime.

Then Paul opens his mouth, and he says, without explanation, "No way."

Paul doesn't tell anybody what he's really thinking, which is this: I'd rather live six months being locked up somewhere with some degree of honor, than to be dishonorable and walk freely in the streets.

Those are Paul's words. Those are Paul's thoughts. But how could you say it out loud without sounding foolish?

There's not much left to say, and a few minutes later, Paul and his lawyer leave the building. They ride down in the elevator to the lobby, past the metal detector, out the doors to the plaza strewn with pigeon droppings.

The indictment has eleven counts. Each count is punishable by at least five years in jail. Each count carries a maximum fine of $250,000.

The first count the grand jury charges is conspiracy. The indictment charges that all of the boys conspired to "gain access to and control of computer systems in order to enhance their image and prestige among computer hackers," a violation of Title 18 of the U.S. Criminal Code.

The other counts charge some of the individual boys with unauthorized access to computers (switches and Tymnet), possession of unauthorized access devices (long-distance calling card numbers), four counts of interception of electronic communications (snatching IDs and passwords from Tymnet files), and four counts of wire fraud (stealing the use of New York University's phone lines to make phone calls to El Paso and Seattle). The indictment is full of fancy language that was really saying only one thing: You could go to jail. So could you and you and you and you. For a long time.

Paul could get five years.

Eli could get five years.

Mark could get ten years.

Julio could get thirty-five years.

John could get forty-five years.

The case is so big, so sensational, so groundbreaking that the U.S. Attorney himself calls a press conference in the lobby at St. Andrews Plaza. He wants to announce the indictment to the media. It's a little off-putting, the rows of folding chairs hastily arranged with their backs to the metal detector and the bullet-proof U.S. marshal's booth. A stream of New York's finest—the press corps, that is—slouches in and starts bitching for handouts.

The indictment is a twenty-three-page document dense with facts, counts, and legalese. The press release that explains what the indictment is trying to say is eight pages long. And then there are charts that Secret Service Agent Rick Harris arranges on an easel. This was before Ross Perot, remember, and the charts are a novel idea. Among other things, the charts explain what a switch is.

The basic point the prosecution is trying to get across is the national scope of the computer intrusions.

"This is the crime of the future," says U.S. Attorney Otto Obermaier, a tall, patrician man in a dark suit. He points a finger to underscore his distaste for computer crimes. "The message that ought to be delivered from this indictment is that this kind of conduct will not be tolerated."

These five boys were being charged with the most widespread intrusions of the nation's largest and most sensitive computer systems ever recorded. The government has decided to make an example of these teenagers from the outer boroughs. The message, which is what Obermaier calls it, is zero tolerance. If you're a hacker, thinking of following in the footsteps of the Masters of Deception, think again.

There is hardly room behind the podium for all the lawmen trying to get a piece of this one. There are the prosecutors. Then there are agents from the FBI and the Secret Service, and there are men from the U.S. Justice Department's computer crime unit.

Obermaier tells the press corps all about the crimes. He tells them the boys' intrusions have cost companies thousands of dollars in security personnel salaries and lost processing time. But he doesn't tell them that the dangerous hackers are, in effect, just a bunch of teenage boys who got to be friends because they shared a hobby. He doesn't mention the long nights in Eli's bedroom, doesn't talk about any of their inside jokes, doesn't say "Plik" once. After Obermaier finishes his speech, the questioning does not go the way you'd expect.

"The indictment says they sold TRW credit reports. So how much did they get?" asks a reporter in the front row.

"Several hundred dollars." Obermaier declines to be more specific.

"Where do the damages come in?" someone asks.

Fishbein explains that the phone companies and Tymnet had to spend a lot of money to fix security holes.

"Isn't that like charging the burglar for the bars you buy to put up on your windows?" someone else asks. It's fair game, Fishbein says, because of the huge number of billable hours the victim companies burned patching up after the kids were through there.

The reporter from the *New York Post* shrugs. His newspaper is a tabloid, the feistiest in town, and he came here in the misguided hope that this would be a hot story. He's been at this job for a long time. He's covered organized crime and organized crime doesn't look like this. He's covered conspiracies, and they don't act like this. These are just kids. There are no bodies, there is no sex, and the crime, if there is one, will be very hard to explain in the small amount of space he'll get.

After the reporters file out, the room is pretty empty. It's just a dingy lobby again. Except for one thing. In the back, way behind the last row of folding chairs, stands Mark.

He's the defendant, and here he is listening to the prosecutors announce his capture. Rarely does the accused appear at the prosecutors' media briefing.

But here he is, the ringleader. A cameraman from one of the TV stations hugs him on the way out.

The story makes NBC's national newscast at six-thirty, with Tom Brokaw. It's the piece Brokaw ends with—a forward-tilt, something-to-think-about piece. There's a headshot of Fishbein speaking to the press and a close-up of Mark teaching a computer class at the New School for Social Research in Greenwich Village. Another shot focuses on Paul, sitting in the first row in the classroom and identified, in a voice-over, as the Scorpion. The piece closes with Mark, talking to the NBC reporter, on a dark and rainy street. Mark says that he's definitely the victim of political persecution.

SIXTEEN

There will never be a trial.

So a lot of the arguments that might have been made about whether what the boys did with their computers was truly wrong would never be aired. A lot of murky issues would never be tested. For instance, was it really against the law to possess, on a floppy disk you carry around at school, the phrase "PHIB*SUX"?

There would never be a jury to decide whether publicizing the credit history of your rival's mom was crime enough to go to prison for a year. Or whether crashing The Learning Link and calling its operators "you turkeys" was crime enough to earn a jail sentence.

If there had been a trial, well, then the MOD conspiracy case could have set a precedent for the entire country, could have established a benchmark by which the government could track down other so-called computer criminals. But there wouldn't be a trial because all the MOD boys had pled guilty over the course of the year. They all had given up. What was left after that? It was up to the prosecutors to recommend a certain sentence, and up to the judge to mete it out.

That's how justice works.

From the day of Otto Obermaier's press conference, the case looks unbeatable to the defense lawyers. The government sends them copies of the evidence. Pages and pages of wiretapped conversations in which John and Julio discussed the intricacies, passwords, and logins, of proprietary corporate systems. Sometimes Mark was on the wiretaps, explaining precisely how to defeat Tymnet's security. A few times, Eli was recorded, too, asking Julio to use TRWNET to look up confidential credit reports. Federal law says it's illegal to break into private computer systems, and here the government had thousands of pages of notes, written in the boys' own hands, that incriminate them. Passwords and logins for phone company computers. Long-distance access codes. And in case the defense attorneys want to argue that simple *possession* of this information was no crime, well, then the government has hundreds of pages of printouts from Dial Number Recorders that show precisely which New York Telephone computers were dialed from the boys' homes.

And do you want to talk about *intent?* Do you want to make the argument that these were just poor scientists, and the world was their laboratory? Then try explaining Julio and John taking money from Morty. Try explaining that phone call where Julio said to Mark, *"I would, like, crash everything."*

The facts trouble Mitch Kapor and John Perry Barlow, whose Electronic Frontier Foundation has been monitoring the progression of the MOD boys' case since the very beginning. They wonder whether the government has taken to picking on defenseless teenagers. One day in 1992, the EFF's staff lawyer (yes, there's a staff now, and an office in Cambridge) comes to New York City to investigate.

The EFF lawyer is named Mike Godwin. He looks over all the government's evidence against Mark, and he consults with Mark's lawyer, Larry Schoenbach. Godwin tells Mitch that the foundation should become more involved in the case, and that if Mark gets up on the witness stand, he might be able to convince a jury that what the boys did wasn't enough to land anybody in jail. Godwin, who once was editor of the daily newspaper at the University of Texas at Austin, has friends among the reporters at *Newsweek* and other national publications. He lays out a sweeping media campaign that the EFF could wage.

But Godwin's boss, Mitch, isn't so sure. For one thing, he doesn't like the idea that Mark kept hacking, even after his first bust. And there's another, bigger reason that the MOD case looks unappealing.

The political world was changing as fast as the online world was in the last half of 1992. Bill Clinton might be headed for the White House, and his would-be vice president, Al Gore, has promised to do everything he can to build what Gore has dubbed a "national information superhighway." The White House will have a lot to say about government's role in policing the new electronic frontier, and the EFF wanted to be in a position to influence the Administration's gurus. People have warned Mitch that at this critical juncture in the young organization's development, the EFF should avoid being labeled a "hacker defense fund." You didn't get anybody's ear in the Capitol with a moniker like that. The EFF's board of directors finally voted to move the organization's headquarters from Boston down to Washington, D.C. If you were an adult, and watching closely, you'd see the message in this. There was more important work to be done than fighting a losing battle on behalf of a bunch of New York City kids who should have known better.

But what the EFF conveys to Mark is: Do what you think you have to do, and we'll help if we can.

Mark doesn't know what to do.

One day in November, he meets Schoenbach at his office on the East Side of Manhattan. For once, Mark arrives early. He wears baggy jeans and a pristine white turtleneck, an outfit complemented by a gold hoop in his left ear and his signature bandanna, which fits neatly over the ring of hair he recently shaved close. He and his lawyer go to lunch at City Luck, a Chinese restaurant around the corner. Mark orders skewered steak—steak on a stick, he calls it—a Coke, and egg drop soup. He does not brave the chopsticks.

Schoenbach and Mark discuss the upcoming meeting with the prosecutors to take place three days from now. Mark's allergies bother him, as usual. He sniffs and sneezes and blows his nose as Schoenbach explains that Fishbein is about to offer a deal to all the MOD defendants. Schoenbach figures that in a case like this, if the defendants plead guilty and obviate the need for a lengthy and costly trial, then the prosecutors could recommend lenient sentences, maybe even no jail time. The MOD boys could all go on with their lives. The downside is that all the boys will be convicted felons. They can't hold office or vote for the rest of their lives.

It's a tough choice for Mark, especially since he doesn't believe he did anything wrong.

Schoenbach advises Mark to take the plea. Now, this advice is worth as much as any lawyer's in this country. Schoenbach has years of experience navigating the federal court system, representing high-profile clients accused of crimes so heinous they don't bear repeating. Terrorists accused of bombing U.S. targets. Alleged Mafia functionaries. Accused murderers. Although Larry has a fancy East Side office (the trompe l'oeil bookshelves in his lobby make the place as chic as any space in Manhattan) he also defends a lot of very poor clients. He's on the federal court system's roster of court-appointed attorneys. Unlike the public defenders who work in state court, these federal defense attorneys are actually guaranteed a decent hourly fee, courtesy of the U.S. court system. It's not bad money, so the job attracts quite a few private practice attorneys who wouldn't normally be attainable. And that's how clients like alleged terrorists and suspected members of Chinese extortion gangs get an attorney like Schoenbach.

Mark is Schoenbach's first client accused of computer conspiracy, and at first Schoenbach may have been a little befuddled by all this talk of Tymnet access, kilobytes, and hacker "meetings" to plan the conspiracy.

But back in August Schoenbach went to see the hackers gather for their *2600* conflagration, and what he saw of the hackers' culture—essentially, a bunch of wispy-moustached boys cavorting awkwardly and shyly in the indoor plaza—convinced him that if you cut through all the technical crap, the case was pretty simple. A bunch of teenagers were trespassing. They got caught. And for this the government wants to label them felons.

Schoenbach doesn't think it's right. But he knows the law, and the law says that what Mark and his friends are accused of doing is illegal. That's the law, and unless you want to argue that the law's wrong, you don't have much of a defense.

He tells all this to Mark. But Mark is trying to navigate an unfamiliar and terrifying terrain, and he can't pull back far enough to see the contour of the landscape. He reminds Schoenbach that nothing really happened the last time he pleaded guilty. Just community service. He hears, but doesn't hear, Schoenbach's warnings that the judge could be harder on a second-time offender. He hears, but doesn't hear, his friends' imprecations to stand united. He plays with his food in the restaurant, he orders a second Coke.

Mike Godwin calls Mark on the phone one night. He tells him that he knows that Schoenbach thinks Mark should plead guilty. He tells Mark that he will stand by him no matter what decision he makes.

At the meeting with Fishbein, the defense attorneys asked whether the prosecutor would recommend a lenient sentence—maybe no jail time at all—if all five boys agreed to plead together. Wrap up the whole case at once, you know? Easier for the prosecution. Fishbein checked with his bosses, and said, well, okay.

But in the meantime, Mark has changed his mind. He refuses to plead. His friends try to convince him otherwise, but he will have none of it.

By this time, all the boys know where they stand. Each has made a trip, accompanied by an attorney, to Fishbein's office. Fishbein has laid out the evidence. Here's what we have on you. The boys sat stony and implacable, taking it in, thinking this: I'm going to jail. *Oh shit, I am really going to jail!*

The facts are glaring to John and Julio. They're charged with the most counts, they're facing the most prison time. After all, John and Julio actually *sold* TRW accounts to Morty Rosenfeld. The government could argue, pretty convincingly, that John and Julio were hacking for profit. They wanted to make money off their crimes. It doesn't look good. John and Julio are likely to get stiffer sentences than the other boys if this case goes to trial and they lose. Both boys talk to their lawyers about the situation. They will plead guilty on December 2.

They arrive at the federal courthouse, a broad wedding cake of a building, looking nervous. John wears a Giants sweatshirt, but carries a white dress shirt still protected by a plastic dry cleaner's bag. Julio wears a suit.

As part of his guilty plea, John submits a written statement:

> I was part of a group called MOD. The members of the group exchanged information, including passwords so that we

could gain access to computer systems which we were not
authorized to access. I got passwords by monitoring Tymnet;
calling phone company employees and pretending to be
computer technicians; using computer programs designed to
steal passwords.

It's not exactly in John's own words, but he and his attorney have
written it out, and now it's court exhibit 3 in the government's case
against him.

I acknowledge that I and others planned to share passwords and transmitted information across state boundaries by
modem or telephone lines and by doing so, obtained the
monetary value of the use of the systems I would otherwise
have had to pay for.

What?
In other words, he made some free calls. Who knows how much
those calls were worth. The government can't even say.
"I apologize for my actions and am very sorry for the trouble I
have caused to all concerned," the statement concludes.
The judge accepts John's plea.
Then Julio stands up and admits that he broke the law: "There
was a phone call I received on November fifth. John called me up and
gave me passwords."
That's all there is to it. They will be sentenced another day. John and
Julio leave the courthouse, walking past a vendor selling umbrellas. The
vendor yells to them, "Snow's coming. Get your umbrella. Snow tonight, tomorrow, and the next day."

From that day on, the Masters of Deception no longer existed.
Word leaked out that Julio had agreed to cooperate with the government, to testify against Mark at his trial. Julio was angry that Mark
had refused a deal. Fishbein won't confirm that Julio was
cooperating, but the hackers all saw through that. Paul no longer
spoke to Julio. Mark no longer spoke to Julio, and refers to him as
"Julio the Rat." Eli tried to make peace between the former friends,
but it was too late.

The five boys weren't standing together anymore. They were on their own.

Mark didn't talk much more about his decision to go to trial. He says he's innocent, of course. What did he steal? What did he break? The EFF will help him, of course. He'd pull a *Miracle on 34th Street* defense: Yes, I did all that stuff, but where was the harm? All it took was one juror to understand that, and there would never be a conviction. Still, the pending trial upset Mark's stomach. The tension was making him physically ill.

The U.S. Attorney's office has to prepare a full-blown trial, prepare dozens of witnesses to testify, organize rooms and rooms of evidence into a coherent argument that will convince a jury. A second assistant U.S. Attorney, Geoffrey Berman, gets assigned to the case with Fishbein. Together, Berman and Fishbein start to marshal their resources. Fred Staples spends hours and hours downtown, helping to decipher the lines of numbers and times and dates that the DNRs have generated. It's not as easy as it might seem, because if a hacker first called an ITT dialup, then used an ITT PIN number to place a call, then dialed a ten-digit long-distance number, all those figures would be run together in a long stream, just as they were originally dialed. So Fred helps sort it out.

Fishbein and Berman make a list of potential witnesses. They interview staff at The Learning Link. They interview the system administrator of Southwestern Bell's C-SCANS computers. They talk to officials from all the other phone companies and businesses who were victimized. They prepare elaborate charts and graphics for the jury—to explain the context of high-tech computer crimes. The other hackers hear that Berman has even tracked down Pumpkin Pete and convinced him to testify.

The first skirmish is a pre-trial hearing that stretches over a two-day period in March of 1993. The purpose of the hearing is to get a ruling from the judge about the admissibility of certain portions of the evidence that the government has gathered. Lawyers for the three remaining defendants try to convince the judge, Richard Owen, to throw out much of the evidence that the government has gathered. The defense attorneys argue that the information from the DNRs was obtained illegally. They argue that as far back as the summer of 1989, the phone company was acting as a de facto agent of the government, gathering information and funneling it straight to the Secret Service.

Because the phone company had been working with the Secret Service, the DNRs should not have been placed on the boys' phones without a court order, they argue.

But then Fishbein stands up to refute the argument: "The statute says that DNRs can be installed to find out about computer intrusions of their systems. New York Telephone was the *victim.* The New York Telephone investigation began well before the Secret Service investigation, and it was done purely to detect fraud and abuse of the system."

Judge Owen sides with the prosecution. He appears to be exasperated by the defendants. Owen likens the phone company to a private citizen who has been robbed of property.

"If a private citizen believes that his stolen property is in an apartment, and he busts in and gets it, and turns it over to the government, then his actions aren't at the behest of the government," Owen says.

Owen's ruling is a blow to the defense, because it means that every bit of information from the DNRs—that damning stream of numbers, now all typed out in neat chart form—can be introduced to a jury.

Attorneys for the defendants also argue a second point, that statements made by Eli and Paul to Secret Service agents should be suppressed, because they were obtained illegally. Paul testifies that he was intimidated by agents who searched his dorm room in 1990, and days later he was coerced into giving a statement about his hacking activities. Owen is skeptical.

"You're not shy," the judge tells Paul, who, truth be told, was blushing mightily at the moment. "You've got a title of Scorpion. There's a little fellow with a big stinger in his tail who gave you that."

Over the years, Owen has developed a reputation as a tough-talking, pro-prosecution judge. He is easily irritated by defense attorneys' assertions that the authorities violate the civil rights of virtually every defendant who stands in his courtroom. Owen has handled too many high-profile cases to list them all, but a couple of years before the MOD boys set foot in his courtroom, Owen sentenced mobsters Tony "Ducks" Corallo, Anthony Salerno, and Carmine Persico to a hundred years each of jail time. He had also presided over the enormously complicated racketeering and fraud trial of a lawyer accused of taking payoffs from the notorious Bronx defense contractor, Wedtech.

The boys standing before Owen's bench are more of the same—defendants accused of breaking federal law.

Soon after Owen ruled against the defendants' efforts to limit evidence against them, Paul and Eli decide to plead guilty. "It's pretty clear, the way things are going, that this is the only choice," Paul says.

At his sentencing, Paul stands before the judge, wearing a suit. A rubber band holds his blond hair in a ponytail that hangs halfway to his waist. "I realize I broke the law. . . . I didn't do it to make money or hurt anyone," he says, apologizing. "A lot of it was for intelligent curiosity, to see how computers were operating."

Paul asks for leniency. "I ask your honor to give me a chance to prove I can offer something to society."

The judge looks sternly at him. "One problem I have with all the young people in this case is you're all so bright. That's really what distresses the deuce out of me."

Paul and Eli each get a sentence of six months in a federal penitentiary, followed by six months' home detention.

Owen sentences John, too.

On a somber, rainy day in the summer of 1993, with his mother, stoic, sitting on a hard wooden bench behind him, John stands in a charcoal gray suit before Owen. He says, "I'm real sorry for the things I've done."

John says that he's grown up a lot in the past year and a half since the raids. He's been having a lot of fun, actually, working as a standup comedian in local clubs. He's also in college now, a film major at Brooklyn College. Filmmaking has become his new obsession. He's been making films on the subways and everyone loves them. The films are funny and they make people laugh. In fact, some of his professors have written to the judge, asking for leniency, because John shows such promise.

"I want to do a film project," John tells the judge. He wants to do an anti-hacking, public-service-type announcement. Put his various talents to good use.

The judge is impassive. John clears his throat and continues. "I've always tried to please my mother and I've let her down. . . . I guess I'm rambling. . . . My mother stood behind me. I've accepted completely that I would have to face jail time for what I've done."

Owen listens, then sits silent for some minutes. Then he says: "He went to the finest honors high school this city has to offer. He was in the swim with some of the brightest men and women. . . . A kid with computer gifts can get out of Stuyvesant and go to any college he wants to. He had before him the opportunity to go absolutely to the limit." The judge looks at John. "You had the world practically given to you."

Who in the courtroom besides John and his mother, Larraine, chokes on that weird statement?

Owen takes the world away. John won't be going back to college in the fall, because the judge sentences him to a year in jail, followed by three years of supervised release and two hundred hours of community service. Oh yes, and a fine of $50.

Then Owen looks over at John's lawyer and asks, "When does your client want to surrender?"

So now Mark was the last holdout. Meanwhile, Julio's sentencing was delayed. Everyone figured that was because Julio had agreed to testify against Mark at trial and would subsequently be rewarded with a suspended sentence. Julio was making regular trips to the U.S. Attorney's office, meeting with Fishbein and Berman, helping them build a case against Mark. Over and over again, Julio reviewed the evidence and the questions he'd be asked on the stand during Mark's trial. He would be the government's star witness. Mark taught me everything I know. . . .

Mark started to prepare for trial, during long meetings with Larry Schoenbach and Schoenbach's partner. He is becoming intimately familiar with the mountain of evidence against him. He is coming to believe that his chances of an acquittal are next to nothing.

And yet, Mark believed he had taken a position that he couldn't back away from.

The EFF was no longer actively involved in his case—a chastened Godwin briefly lost his job partly because he pushed his bosses for a commitment to the case. Godwin backed off after he was reinstated as staff counsel. Barlow felt bad about the whole thing. He believed that the mixed signals the EFF had sent had turned the MOD case into a bloody casualty of the EFF's struggle to define itself as an organization. He viewed the foundation and its founders as deeply divided (he himself has never known whether he wanted to be a Republican or an outlaw, and he wondered if Mitch would rather be a guru or a CEO). Barlow thought that by backing away from the case, the EFF might be betraying its own roots. But by now, events were beginning to overtake everyone.

Mark's lackluster participation has soured Larry Schoenbach as well. Mark was late to meetings. Mark skipped meetings. Mark was taciturn and sullen. If this was how he behaved when dealing with his lawyer, how would he look to a jury when he took the stand in his own defense?

The trial was scheduled to begin July 5.

During the last week in June, Fishbein and Berman put the finishing touches to their case. Berman rehearsed his opening statement to the jury.

Then, on Thursday, five days before the start of trial, Mark's lawyer called Fishbein and said, "We really should talk about a plea."

The next morning, Friday, Mark stood in court and echoed the words of his friends: "I plead guilty, your honor."

Why plead guilty now? Schoenbach said it made no sense. Had Mark gone to trial and lost, it is unlikely he would have earned more time in jail than he did by taking a plea so late in the game.

But Mark has told people that by pleading on the eve of trial, he got something none of the other defendants did: a complete understanding of what his defense would have been. He saw every last plank of support laid out.

And in the end, what he'd seen scared him.

AFTER

The hackers who arrive at the Citicorp building on the first Friday of February wear heavy hiking boots, thick-soled turf crunchers, and kicked-around, black-leather shoes that lace up, up, up their calves. Of course, if it were July, they wouldn't be dressed any differently. This is the uniform.

It is 1994 now, and tonight's meeting is the first gathering in nearly five years at which no one from MOD is present. Dozens of hackers are here, ranging in age from fourteen to forty, far more muffle-jacketed attendees than in the days of early 1989.

The world has changed since that heady time when Mark and Paul and Eli and John and Julio all somehow found each other, all somehow coalesced.

In fact, it's as if the rest of the world has caught up. What the MOD boys did for fun—recreationally cruising across continents of wires—has become a national pastime. "Net surfing" is a bigger fad than CB radio ever was, and people everywhere are buying their first computers and hooking up to online services that connect them to the world and one another. My mom. Your mom. Everybody's entranced.

So it's no wonder that the new numbers of an eager generation are filling the Citicorp atrium. Tonight, there's a hole where Mark used to be, a spot by the pay phones where he liked to stand patiently while a group of respectful protégés would gather to ask him highly technical questions. Tonight you will not see his familiar blue-and-white bandanna, you will not hear the boom-and-heave of his voice, you will not wonder how many bowls of homemade chicken soup he will consume later downtown. Tonight, Mark is far away.

Mark had arrived at the gates of Pennsylvania's Schuylkill Prison late one night in January, right after a snowstorm, and was whisked inside before his friends could say good-bye. He started serving a year-long sentence, the longest stint in jail that any of the MOD boys would serve. The sentencing judge said that Mark by his actions chose to be a messenger for the hacking community. And so the judge had said he had no recourse but to send a message back.

Hundreds of people had sent letters to the judge, urging him to be lenient. The letters came from an electronic community that had flourished with Mark's help. The community is called Echo, which stands for East Coast Hang Out. Echo was a local New York City bulletin board before Mark was hired as a system administrator, a job he took while waiting to be sentenced. Mark quickly connected Echo to the Internet, the worldwide collection of more than fifteen thousand interconnected computer networks.

The world of the Internet, for newcomers, is a fabulously confusing place. In other words, a perfect place for Mark Abene to de-mystify. And the members of Echo—lawyers, doctors, school-teachers, writers—came to rely on Mark as their instructor and travel guide. He found, as lucky adults often do, creative work that consumed him. And that paid him a wage. His job will be waiting when he gets out of jail.

On the day Mark drove off to prison, many of his friends on Echo add the name "Optik" to their online names. Someone printed up buttons that say PHREE PHIBER OPTIK and *The Village Voice* ran a feature calling Mark the first martyr of the information age. A kind of vigil is held every day, online, while he's in jail. A list is set up on Echo of things Mark needs, and people send him magazines and books and hundreds of letters.

On Echo he had also found a girlfriend. She misses him desperately and every day posts a message on Echo, updating his new friends on his condition and mood in jail. She says she just wants her Phiber back home, sitting in the bathtub with her and catching the Milk Duds she tosses right into his mouth.

* * *

Paul is in jail, too. He's serving six months at a federal prison in Lewisburg, not so far away from Mark. When he gets out, he will also have a job waiting for him, working for some guy who runs a business called The Missing Person's Bureau, which tracks missing people. There were a lot of people who wanted to hire Paul. The computer industry is not afraid of his felony conviction. The industry is filled with rebels, old-line hackers who understand how a brilliant kid could take a wrong turn and still be a brilliant kid. Don't worry about Paul.

John was about to get out of jail, thinner, more muscular, in the best shape he's ever been in his whole life. He thinks about college and asks friends to send him movie reviews. He can't wait to get back to his life in the film studies program he was enrolled in at Brooklyn College. He's got lots of ideas for films he wants to make.

Eli also has a job waiting for him when he gets out of Allenwood, the most famous of federal prisons. He'll program computers for a broadcasting company in Manhattan, working for a guy who has a computer science degree and who thinks Eli has a great amount of potential.

They had all finally managed to learn how things work. They had all grown up and figured out how to fit themselves into a world that wasn't ready for their skills or their curiosity when they were younger.

And Julio? He and Allen (who wasn't even indicted) were the only ones who didn't get jail sentences. The government was so thankful that Julio offered to testify against Mark at trial that he escaped with a suspended sentence. He's been working for his uncle, who runs a dental lab, setting up computers.

The LOD is just a memory, too. Comsec went out of business and Chris Goggans is working for a large computer maker in Austin, Texas, researching advanced wireless networks. He's also assembled thousands of messages from various elite bulletin boards in the 1980s and sells the collection through a company called LODCOM. At the last Ho-Ho Con, he was selling T-shirts, too. They said THE HACKER WAR across the chest and there was a map of the United States depicting Chris's version of major battle sites, mostly Houston and

Austin and New York. On the back it said LOD 1, MOD 0. And there
was a quote, attributed to Corrupt, that Chris said he got off
MODNET:

"It's not just winning that counts but making sure that everyone
else loses."

Would anyone from MOD ever again come to the hackers'
meetings at Citicorp?

Tonight, a crowd is pecking and jabbing at the phones, right where
John used to stand when he sweet-talked the operators. Nearby, a dark-
jacketed youth sought advice from another wearing a Mets cap
backward: "You just go to the speakers and you move the phone jack.
It's simple." Kids wearing beepers on their belts and cellular phones in
their pockets greet one another. They pull printouts from their
knapsacks to pass around: they cluster around a copy of the new *2600*
magazine. The latest issue features stories about the MOD boys'
plights, and on the cover is a rag doll stabbed in the heart by a dagger.
The shaft is labeled: BERMAN.

This is the next generation of hackers. MOD is the stuff of legends
now, and the wide-open stretches of cyberspace that Mark and his
friends had roamed have been fenced off by corporate owners. Still, the
hackers flourish. Are any new conspiracies fomenting today?

Well, there's the fuss about the Internet. Supposedly, some new
group of hackers who call themselves The Posse have been breaking in
to the Internet, which has become by default the world's Information
Superhighway. The number of Internet users is growing faster than the
population of turn-of-the-century New York City. University networks.
State networks. Private networks. Library networks. Twenty million
people use the Internet every day, and at the present rate of increase,
the entire population of the world will be on the Net by the year 2003.

The Net has virtually no security. It makes Tymnet look like a
fortress. So it wasn't so hard for The Posse to capture thousands of
users' passwords before system administrators noticed the break-ins.
Nobody knows what this new generation of hackers has in mind, but it
sounds ominous. Plus, the modus operandi sounds familiar, so familiar,
that the FBI paid a visit to Allen recently. Of course, Allen wasn't
charged with anything. Just an informational visit, you understand.

Tonight, there's a bunch of kids crowded around one of the little
cafe tables in the indoor plaza, hunched over some inexplicable pile

of wires rising like a crazy tornado from the tabletop, cords spilling over the edge, winding carelessly through the legs of a chair. The kids are intent on their work, whatever their work may be. One thing's noticeably different from 1989. Not all these kids are white. In fact, look around, and you'll see that half of the new hackers hail from Jamaica (the island, not the Queens neighborhood), from the projects, from Colombia. They're first-generation settlers, they're the children of immigrants, the kids who moved into the neighborhoods where the MOD boys grew up.

There's a nearly blind hacker, you can tell that his eyes don't focus because they aren't even the same size, aren't even open the same amount. He walks up to any dark shape he sees, and issues a greeting: "Are you new here? I don't remember you from last month." Maniac, a supremely wan ponytailed hacker from Brooklyn who wears thick black wraparound lenses for privacy, skulks the perimeter, swathed in a long coat. Razor, an ebullient fourteen-year-old whose social skills are remarkable in this group simply because they exist, flits from hacker to hacker, shaking hands, introducing himself, beaming a jowly grin. The Twitching Hacker, the nervous one who used to have a whole list of questions for Mark on topics like why the tone skipped a click when you dialed a 9, looks a little forlorn. So does Eric Corley, who carries with him the memory of dropping off Mark at the prison gates. Eric, who is thinner than ever, is dressed all in black, T-shirt and open-necked shirt and jeans. He looks older now, and steadies his hand against his chin when he talks. He carries a scanner, and invites people to "hear a conversation in Russian."

And into this fray, this wildly discombobulated massing of alienated adolescence, walks Barlow.

It's a quiet entrance, just a man in a black, belted leather jacket and throat scarf ambling into the center of the activity. And stopping. Barlow stands there, hands in pockets, eyes all lit up at the wonder of this new experience.

John Perry Barlow lives in New York City now, where he's the ambassador from the Electronic Frontier Foundation. Sometimes he goes on TV. Sometimes he writes articles. The foundation has become a power player in Washington, helping draft the blueprint for Vice President Al Gore's vision of cyberspace. Some people say the foundation has abandoned the little guy to embrace big ideas instead.

Tonight, for the first time since he moved back east, for the first time since the December evening four years ago when he ran across two electronic explorers named Phiber Optik and Acid Phreak, Barlow has come to a *2600* meeting, but somehow Barlow arrived too late.

His presence does not go unnoticed. Some of the hackers recognize him right away. "Barlow—that's a trip," says Corley and comes to greet him.

Razor, who is about to launch his own commercial Internet gateway, sees him, too, and hustles over to shake hands. "Hey, I met you once, when you were with Bruce Sterling," says Razor, wedging himself in next to Corley. Unaware of what all the fuss is about, Maniac wanders up, gets introduced, half-hears Barlow's name, wanders off. He doesn't know who he's met, not yet, but it will only be minutes before he realizes. Barlow. The news whizzes around the atrium at the speed of thought, as Mark would have said.

"Did I miss the meeting?" Barlow asks Corley, watching the milling clumps of boys. They are the picture of entropy, of disorganization, of isolated growing pains and undeveloped social skills.

"No, this is the meeting."

Barlow's eyes sweep the indoor plaza, taking it all in. The Asian kid sitting cross-legged on the low wall, alone and reading *2600*. The blind guy. The table covered with wires. Maniac's eyeglasses. Razor's grin. German tourists standing in line to use a pay phone, puzzled by the wait. If he notices the absence of the most famous of hackers, he does not comment on it. If he thinks about how he once offered support to them, he does not say so.

"This *is* the meeting," Barlow says. "I get it. I get it." And he laughs a huge, deep laugh. "Of course this is the meeting."

This is the conspiracy.